# Springer Theses

Recognizing Outstanding Ph.D. Research

For further volumes:
http://www.springer.com/series/8790

## Aims and Scope

The series "Springer Theses" brings together a selection of the very best Ph.D. theses from around the world and across the physical sciences. Nominated and endorsed by two recognized specialists, each published volume has been selected for its scientific excellence and the high impact of its contents for the pertinent field of research. For greater accessibility to non-specialists, the published versions include an extended introduction, as well as a foreword by the student's supervisor explaining the special relevance of the work for the field. As a whole, the series will provide a valuable resource both for newcomers to the research fields described, and for other scientists seeking detailed background information on special questions. Finally, it provides an accredited documentation of the valuable contributions made by today's younger generation of scientists.

## Theses are accepted into the series by invited nomination only and must fulfill all of the following criteria

- They must be written in good English.
- The topic should fall within the confines of Chemistry, Physics, Earth Sciences, Engineering and related interdisciplinary fields such as Materials, Nanoscience, Chemical Engineering, Complex Systems and Biophysics.
- The work reported in the thesis must represent a significant scientific advance.
- If the thesis includes previously published material, permission to reproduce this must be gained from the respective copyright holder.
- They must have been examined and passed during the 12 months prior to nomination.
- Each thesis should include a foreword by the supervisor outlining the significance of its content.
- The theses should have a clearly defined structure including an introduction accessible to scientists not expert in that particular field.

Takuya Kanazawa

# Dirac Spectra in Dense QCD

Doctoral Thesis accepted by
The University of Tokyo, Japan

Springer

*Author*
Dr. Takuya Kanazawa
University of Regensburg
Regensburg
Germany

*Supervisor*
Prof. Tetsuo Hatsuda
Theoretical Research Division
Nishina Center, RIKEN
Wako
Japan

ISSN 2190-5053 ISSN 2190-5061 (electronic)
ISBN 978-4-431-54164-6 ISBN 978-4-431-54165-3 (eBook)
DOI 10.1007/978-4-431-54165-3
Springer Tokyo Heidelberg New York Dordrecht London

Library of Congress Control Number: 2012943356



Printed on acid-free paper

Springer is part of Springer Science+Business Media (www.springer.com)

**Parts of this thesis have been published in the following journal articles:**

- T. Kanazawa, T. Wettig and N. Yamamoto, "Singular values of the Dirac operator in dense QCD-like theories," *JHEP* **1112** (2011) 007
- G. Akemann, T. Kanazawa, M. J. Phillips and T. Wettig, "Random matrix theory of unquenched two-colour QCD with nonzero chemical potential," *JHEP* **1103** (2011) 066
- T. Kanazawa, T. Wettig and N. Yamamoto, "Exact results for two-color QCD at low and high density," *PoS* **LAT2010** (2010) 219
- T. Kanazawa, T. Wettig and N. Yamamoto, "Chiral random matrix theory for two-color QCD at high density," *Phys. Rev.* **D81** (2010) 081701
- T. Kanazawa, T. Wettig and N. Yamamoto, "Chiral Lagrangian and spectral sum rules for two-color QCD at high density," *PoS* **LAT2009** (2009) 195
- T. Kanazawa, T. Wettig and N. Yamamoto, "Chiral Lagrangian and spectral sum rules for dense two-color QCD," *JHEP* **0908** (2009) 003
- N. Yamamoto and T. Kanazawa, "Dense QCD in a Finite Volume," *Phys. Rev. Lett.* **103** (2009) 032001

# Supervisor's Foreword

Physics of the strong interaction at finite temperature and/or baryon density is one of the most interesting topics in modern nuclear and particle physics. At temperatures greater than $10^{12}$ K, nuclei dissolve into a plasma of quarks and gluons, known as quark–gluon plasma: our universe started from such an extreme state of matter, and experimental efforts are currently being undertaken to create the quark–gluon plasma by using the relativistic heavy-ion collisions at the RHIC (Relativistic Heavy Ion Collider) and at LHC (Large Hadron Collider). For baryon densities greater than $10^{12}$ kg/cm$^3$, nuclei dissolve into a degenerate Fermi system of quarks, so-called quark matter. This exotic matter may be formed in the central core of the neutron stars.

The theoretical basis for describing the quark–gluon plasma and the quark matter is provided by quantum chromodynamics (QCD), specifically the color SU(3) gauge theory for the strong interaction. Due to its highly non-perturbative nature, solving QCD analytically is one of the most challenging problems in modern mathematical physics. On the other hand, Monte Carlo simulations of QCD formulated on a space–time lattice, referred to as lattice QCD, have become a powerful tool for solving QCD numerically. However, lattice QCD simulations at finite baryon density are extremely difficult because of the notorious "sign problem" caused by the complex QCD action at finite baryon chemical potential; this sign problem has been preventing us from making progress in "dense QCD" for many years.

The present thesis by Dr. Takuya Kanazawa introduces a novel way to derive non-perturbative and accurate results in dense QCD. Taking two-color QCD (color SU(2) gauge theory) as a primary example, Dr. Kanazawa has studied the eigenvalue distribution of the Dirac operator by formulating and solving a new non-Hermitian chiral random matrix theory (ChRMT). In particular, he shows that the ChRMT with weak non-Hermiticity can be mapped to two-color QCD with small chemical potential, and the ChRMT with strong non-Hermiticity can be mapped to the BCS regime of two-color QCD at high density. The spontaneous symmetry-breaking pattern of dense two-color QCD plays a key role in proving such correspondence. The approach developed by Dr. Kanazawa and his colleagues offers a theoretical framework not only for a deeper understanding of the sign problem in QCD and QCD-like theories but also for new insights into QCD phases from low density to high density. Also, the present thesis contains a transparent description of dense QCD and its relation to chiral perturbation theory and chiral random matrix theory, making it a good introductory monograph for students and researchers who are interested in this field.

It is with the greatest pleasure that I introduce Dr. Takuya Kanazawa's work for publication in the Springer Theses series. His work was nominated as an outstanding Physics Ph.D. Thesis of the Fiscal Year 2010 by the Department of Physics, Graduate School of Science, the University of Tokyo.

Tokyo, March 2012 Tetsuo Hatsuda

# Acknowledgments

I would like to express my wholehearted gratitude to Prof. T. Hatsuda for our fruitful discussions on various topics in physics, his continuous encouragement throughout the last five years, and his critical reading of my doctoral thesis. I am also grateful to Prof. T. Wettig for patiently helping me to understand random matrix theories and giving me the opportunity to stay at the Physics Department of the Regensburg University in October 2010. My gratitude also goes to Dr. N. Yamamoto for opening my eyes to the rich physics of dense QCD. I learned numerous important concepts and techniques in quantum field theory from my illuminating discussions with him. Were it not for fruitful collaborations with Prof. Wettig, Prof. G. Akemann, Dr. Yamamoto and M. Phillips, it would have been entirely impossible to complete the research that led to this thesis. I owe most of my knowledge on condensed matter physics to S. Uchino, H. Shimada, T. Kimura, T. Sagawa and N. Sakumichi. My discussions with them have always been a great source of delight and amazement for me. I would like to acknowledge tireless support of all members of the nuclear theory group, especially Prof. S. Sasaki, Prof. T. Hirano, Dr. N. Ishii, A. Rothkopf, Y. Akamatsu and Y. Hirono. I am grateful to Prof. T. Matsui for having taught me the richness and profoundness of theoretical physics when I was an undergraduate student at the Komaba campus. I thank Prof. H. Fujii, N. Tanji and the late Y. Ohnishi of the Komaba nuclear theory group for our valuable discussions in study meetings during my first two years at the graduate school. I sincerely thank Prof. J. Verbaarschot, Prof. T. Yoneya, Prof. Y. Kikukawa, Prof. S. Ejiri, Prof. M. Ueda, Prof. M. Nitta, Prof. M. Tachibana and Prof. K. Fukushima for our stimulating discussions on various topics in physics, including quark confinement, lattice QCD, cold atomic gases, topological solitons, and dense QCD.

Finally I would like to express my deep appreciation to my family for supporting and encouraging me through the hard times and good times alike.

# Contents

1 **Introduction** . . . . . . . . . . 1
References . . . . . . . . . . 5

2 **QCD with Chemical Potential and Matrix Models** . . . . . . . . . . 7
2.1 Phases of QCD . . . . . . . . . . 7
2.1.1 Overview . . . . . . . . . . 7
2.1.2 Color Superconductivity and the CFL Phase . . . . . . . . . . 8
2.2 Two-Color QCD at Low Density . . . . . . . . . . 14
2.2.1 Basics . . . . . . . . . . 14
2.2.2 Why ChPT is Worth Doing . . . . . . . . . . 18
2.2.3 ChPT for Two-Color QCD at Small $\mu$ . . . . . . . . . . 20
2.3 Chiral Random Matrix Theory . . . . . . . . . . 23
2.3.1 Dirac Spectrum and ChPT . . . . . . . . . . 23
2.3.2 The $\varepsilon$-regime of ChPT . . . . . . . . . . 28
2.3.3 Chiral Random Matrix Theory (ChRMT) . . . . . . . . . . 32
2.3.4 ChRMT at $\mu \neq 0$ . . . . . . . . . . 37
References . . . . . . . . . . 41

3 **Dirac Operator in Dense QCD** . . . . . . . . . . 51
3.1 Dense QCD in a Finite Volume . . . . . . . . . . 51
3.2 Dense Two-Color QCD . . . . . . . . . . 57
3.2.1 Symmetry Breaking and Chiral Lagrangian . . . . . . . . . . 57
3.2.2 Finite-Volume Analysis . . . . . . . . . . 63
3.3 ChRMT for Dense Two-Color QCD: Construction . . . . . . . . . . 70
3.4 ChRMT for Dense Two-Color QCD: Solution . . . . . . . . . . 75
3.4.1 Spectral Density at Finite $N$ . . . . . . . . . . 76
3.4.2 The Finite-Volume Partition Function . . . . . . . . . . 79
3.4.3 Microscopic Spectral Density at Maximal Non-Hermiticity: $\hat{\mu} = 1$ . . . . . . . . . . 81
3.4.4 Microscopic Spectral Density at Weak Non-Hermiticity: $N\hat{\mu}^2 \sim \mathcal{O}(1)$ . . . . . . . . . . 83

3.5 Sign Problem in Dense Two-Color QCD and ChRMT . . . . . . . . 87
3.5.1 Structure of the Dirac Spectrum . . . . . . . . . . . . . . . . . . 87
3.5.2 Average Sign Factor . . . . . . . . . . . . . . . . . . . . . . . . . 91
References . . . . . . . . . . . . . . . . . . . . . . . . . . . . . . . . . 96

**4 Three-fold Way at High Density** . . . . . . . . . . . . . . . . . . . . 101
4.1 Introduction . . . . . . . . . . . . . . . . . . . . . . . . . . . . . 101
4.2 QCD at Large Isospin Density ($\beta = 2$) . . . . . . . . . . . . . . . . 102
4.2.1 QCD Side . . . . . . . . . . . . . . . . . . . . . . . . . . . . 102
4.2.2 RMT Side . . . . . . . . . . . . . . . . . . . . . . . . . . . . 107
4.3 QCD with Real Quarks at Large Density ($\beta = 4$) . . . . . . . . . . . 113
4.3.1 QCD Side . . . . . . . . . . . . . . . . . . . . . . . . . . . . 113
4.3.2 RMT Side . . . . . . . . . . . . . . . . . . . . . . . . . . . . 116
References . . . . . . . . . . . . . . . . . . . . . . . . . . . . . . . . . 120

**5 Conclusions** . . . . . . . . . . . . . . . . . . . . . . . . . . . . . . . 123
References . . . . . . . . . . . . . . . . . . . . . . . . . . . . . . . . . 125

**Appendix A** . . . . . . . . . . . . . . . . . . . . . . . . . . . . . . . . 127

# Abbreviations

| | |
|---|---|
| BCS mechanism | Bardeen-Cooper-Schrieffer mechanism |
| CFL phase | Color-flavor-locked phase |
| ChGOE | Chiral Gaussian orthogonal ensemble |
| ChGSE | Chiral Gaussian symplectic ensemble |
| ChGUE | Chiral Gaussian unitary ensemble |
| ChPT | Chiral perturbation theory |
| ChRMT | Chiral random matrix theory |
| ChSB | Chiral symmetry breaking |
| EFT | Effective field theory |
| HDET | High density effective theory |
| NG modes | Nambu–Goldstone modes |
| pqChPT | Partially quenched chiral perturbation theory |
| QCD | Quantum chromodynamics |
| RMT | Random matrix theory |
| SSB | Spontaneous symmetry breaking |

# Chapter 1
# Introduction

It was in 1897 that J. J. Thomson made experiments which led to the discovery of *the electron*: the first elementary particle discovered in nature. After a century we now understand that all visible material in our Universe is composed of a few kinds of elementary particles, called quarks, leptons and gauge bosons. The Standard Model of particle physics is by now firmly established as *the* theory of elementary particles. It is a gauge theory based on the gauge group U(1) × SU(2) × SU(3). While the U(1)×SU(2) sector takes care of the electromagnetism and the weak force, the SU(3) sector, called *Quantum Chromodynamics (QCD)*, describes the strong interactions that act between quarks and gluons. The Lagrangian of QCD is given by[1]

$$\mathcal{L}_{\text{QCD}} = \sum_{f=1}^{N_f} \overline{\psi}_f (i\gamma^\mu D_\mu - m_f)\psi_f - \frac{1}{4} F^a_{\mu\nu} F^{a\mu\nu} \tag{1.1}$$

$$D_\mu \equiv \partial_\mu + igA^a_\mu T^a, \qquad a = 1, 2, \ldots, N_c^2 - 1 \tag{1.2}$$

$$F^a_{\mu\nu} = \partial_\mu A^a_\nu - \partial_\nu A^a_\mu + g f^{abc} A^b_\mu A^c_\nu, \tag{1.3}$$

where $T^a$ and $f^{abc}$ denote the generators and the structure constants of SU($N_c$), respectively. The integers $N_f$ and $N_c$ specify the number of flavors and colors; in reality $N_f = 6$ and $N_c = 3$. Usually we can neglect the contribution of heavier flavors ($c$, $b$, $t$) so that $N_f = 3$ or even 2 is a good approximation to the reality. Despite its simplicity Eq. (1.1) gives rise to incredibly diverse phenomena, such as the emergence of numerous bound states (pions, Kaons, vector mesons, baryons and atomic nuclei), chiral anomaly, asymptotic freedom, gluon condensate, spontaneous chiral symmetry breaking, and color confinement.

The asymptotic freedom [1, 2] is a property that the gauge coupling $g$, which depends on the energy scale under consideration as a result of renormalization, flows to zero at asymptotically high energy. This salient feature of QCD provided us with the opportunity to test QCD by making predictions on high-energy scattering

[1] Electromagnetic interactions are neglected.

T. Kanazawa, *Dirac Spectra in Dense QCD*, Springer Theses,
DOI: 10.1007/978-4-431-54165-3_1, © Springer Japan 2013

processes using perturbation theory and comparing with experiments. On the other hand, $g$ grows large at low energy where weak coupling methods are doomed to failure; phenomena such as bound state formation and symmetry breaking never occur within perturbation theory. Instead, we nowadays carry out massive computer simulations of QCD on a discrete lattice (called *lattice QCD*), and thereby succeeded in revealing various non-perturbative aspects of QCD from first principles. It has to be noted, however, that numerical simulations do not necessarily entail a physical understanding of the subject.

Chiral symmetry and its breaking are one of the central ingredients in low-energy QCD. When all quark masses are tuned off, the Lagrangian Eq. (1.1) is invariant under

$$\mathrm{SU}(N_f)_L \times \mathrm{SU}(N_f)_R \times \mathrm{U}(1)_B \times \mathrm{U}(1)_A, \tag{1.4}$$

where $\mathrm{U}(1)_A$ is explicitly broken by chiral anomaly. In vacuum the chiral condensate $\langle \bar{\psi}_R \psi_L + \bar{\psi}_L \psi_R \rangle \simeq -(240\,\mathrm{MeV})^3$ develops, which breaks the symmetry as $\mathrm{SU}(N_f)_L \times \mathrm{SU}(N_f)_R \to \mathrm{SU}(N_f)_V$. This *spontaneous breaking of chiral symmetry* is still poorly understood from a theoretical perspective owing to the strong coupling nature of QCD, but it has been confirmed in lattice simulations (see e.g., [3]). It gives rise to $N_f^2 - 1$ massless Nambu–Goldstone bosons (pions) and large dynamical mass for quarks, which is independent of the current quark masses $m_f$ in Eq. (1.1). Those light pions dominate the infrared dynamics of QCD and particularly serve as a source of the attractive interaction between nucleons, which in turn leads to the formation of atomic nuclei. The chiral symmetry breaking therefore plays a major role in making our world appear as we see now.

It has been well known since the seminal work of Banks and Casher [4] that the magnitude of the chiral condensate $\langle \bar{\psi}\psi \rangle$ in the chiral limit is linked to the density of near-zero eigenvalues of the Euclidean Dirac operator $\gamma^\mu D_\mu$. That the spectral density indeed tends to a nonzero value near the origin has been explicitly demonstrated in a recent lattice simulation [5], as shown in Fig. 1.1. It is by now established that the way the thermodynamic limit is approached in the low end of the Dirac spectrum is described by *Chiral Random Matrix Theory* (ChRMT) [6]. ChRMT is one of rare examples in QCD where non-perturbative results can be obtained exactly. In ensuing chapters we will review this subject in much greater detail.

The purpose of this thesis is to argue that a similar theoretical scheme can be developed for the Dirac operator in QCD with $N_c = 2$ (called *two-color QCD*) *at large quark number density*. To clarify how unexpected this extension is, let us review qualitative differences between QCD at zero density (vacuum) and QCD at high density:

1. At zero density the strong coupling effect is essential, while at high density the running coupling $g$ reduces to zero owing to the asymptotic freedom, and it is the existence of the Fermi surface that plays a key role in various non-perturbative phenomena of dense QCD,
2. The patterns of spontaneous breaking of global symmetries are totally different in QCD vacuum and at high density, and

3. The eigenvalues of the Euclidean Dirac operator become complex in the presence of nonzero quark chemical potential $\mu$. Consequently the statistical measure of the path integral is no longer positive definite, signaling the notorious sign problem.

Nonetheless it is possible to generalize various theoretical apparatuses originally developed for zero density QCD to the high density BCS regime. Let $D(\mu)$ denote the Euclidean Dirac operator at nonzero chemical potential $\mu$. In this thesis, the following generalizations of seminal works in the literature are completed:

- Chiral Lagrangian at $\mu < \Lambda_{\text{QCD}}$ [7, 8]
  $\Longrightarrow$ Chiral Lagrangian at $\mu \gg \Lambda_{\text{QCD}}$ … Sect. 3.2.1

- Partition function in the $\varepsilon$-regime at $\mu \sim \dfrac{1}{\sqrt{V_4}F_\pi}$ [9, 10]
  $\Longrightarrow$ Partition function in the new $\varepsilon$-regime at $\mu \gg \Lambda_{\text{QCD}}$ … Sect. 3.2.2

- ChRMT for the eigenvalues $\{\lambda_n\}$ of $\mathcal{D}(\mu)$ in the microscopic domain $\left(|\lambda_n| \sim \dfrac{1}{V_4|\langle\bar{\psi}\psi\rangle|}\right)$ at $\mu \sim \dfrac{1}{\sqrt{V_4}F_\pi}$ [11–13]
  $\Longrightarrow$ ChRMT for the eigenvalues $\lambda_n$ of $D(\mu)$ in the new microscopic domain $\left(|\lambda_n| \sim \dfrac{1}{\sqrt{V_4\Delta^2}}\right)$ at $\mu \gg \Lambda_{\text{QCD}}$ … Sects. 3.3 and 3.4

- Analysis of the sign problem in the $\varepsilon$-regime at $\mu \sim \dfrac{1}{\sqrt{V_4}F_\pi}$ [14, 15]
  $\Longrightarrow$ Analysis of the sign problem in the new $\varepsilon$-regime $\mu \gg \Lambda_{\text{QCD}}$ … Sect. 3.5

- The "three-fold way" of ChRMT at $\mu \sim \dfrac{1}{\sqrt{V_4}F_\pi}$ [16]
  $\Longrightarrow$ A new "three-fold way" of ChRMT at $\mu \gg \Lambda_{\text{QCD}}$ … Chap. 4

It is impressive that in the $\varepsilon$-regime so many non-perturbative results can be obtained *exactly*. This tractability is due to the fact that the infinitely many degrees of freedom of quantum fields decouple dynamically in the $\varepsilon$-regime, leaving behind only a finite number of degrees of freedom that can be studied analytically.

Several comments are in order.

### Why Two Colors Instead of Three Colors?

We would like to address a possible criticism: Why $N_c = 2$? Why not work on $N_c = 3$? The reasons will be spelled out in the following chapters and here we present a brief summary of them. First, contrary to $N_c = 3$ QCD which faces a severe sign problem at $\mu \neq 0$, two-color QCD is amenable to lattice simulations at any $\mu$ (as long as the masses are pairwise degenerate). Hence there is a hope that our analytical results can be tested on the lattice directly. Such a comparison will serve as a useful check of simulation algorithms and will prompt further developments in this field. Secondly, two-color QCD and $N_c = 3$ QCD share the important physics of Cooper pairing near the Fermi surface at high density, even though $N_c = 2$ is theoretically much simpler than $N_c = 3$ (e.g., because the quark pair of the former does not break gauge symmetry). Thus one can expect that from a thorough analysis of dense two-color QCD we will learn at least certain universal aspects of dense $N_c = 3$ QCD. And last, but not least, our analysis of dense two-color QCD permits a straightforward generalization to other QCD-like theories, as is elucidated in Chap. 4. In this sense our current treatment is more general than it appears at first sight.

The extension to dense $N_c = 3$ QCD is a highly challenging attractive problem begging for a solution. This is left for future work.

### Matrix Models as a Phenomenological Model of QCD

While this thesis is mainly concerned with ChRMT as an *exact* theory of the QCD Dirac spectrum in the microscopic domain, the ChRMT has been also used as a phenomenological mean-field model of QCD at finite temperature and density. In this context, the universal correspondence of ChRMT to the $\varepsilon$-regime QCD is lost, but we can learn about qualitative features of the QCD phase diagram. The reader interested in this direction is referred to a recent review [17].

### Subsequent Progress

As we will discuss in the main text, the high-density regime and the low-density regime of dense two-color QCD are described by distinct low-energy effective theories, because the patterns of symmetry breaking are different. This observation triggers a question about the physics at intermediate densities. In [18] it was shown, on the basis of the putative BEC-BCS crossover scenario, that the three regimes of dense two-color QCD with high, intermediate, and low density, respectively, are governed by three distinct effective theories, with overlapping domains of validity. The relation between those theories was analyzed in detail and found to be totally consistent with each other. Eventually it was shown that the scheme of the present thesis can be successfully extended to *all densities.*

Furthermore, we also succeeded in extending the universal correspondence between ChRMT and the Dirac spectrum in QCD to a novel correspondence between ChRMT and the *singular value spectrum* of the Dirac operator [18]. The key observation in this work is that the singular value spectrum can be probed through the addition of the *diquark sources* in place of usual quark masses. Important results such as the Banks–Casher relation, the Smilga–Stern relation and the Leutwyler–Smilga sum rules, for the Dirac eigenvalues are now all generalized to the case of the Dirac singular values.

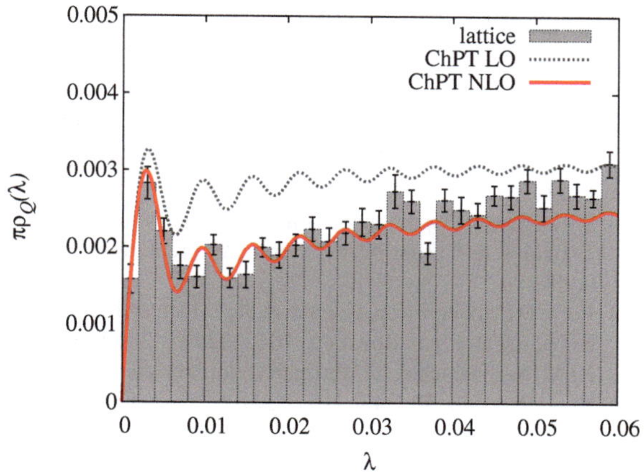


**Fig. 1.1** Spectral density of the Dirac operator measured in a lattice QCD simulation. (Reprinted from [5]. Copyright(2012) with permission from Elsevier)

Since these developments substantially widen and deepen the framework proposed in this thesis, one may view this thesis and [18] as comprising a single integrated research. The reader interested in this thesis is therefore highly recommended to take a look at [18] as well.

This thesis is organized as follows. In Sect. 2.1 we will explain the motivation to look into QCD at nonzero $\mu$ and review past studies on the phase diagram of QCD. In Sects. 2.2 and 2.3 we will review important features of two-color QCD and ChRMT. In Chap. 3 a thorough investigation is given into various properties of the Dirac spectrum in dense two-color QCD: we will construct a low energy effective theory of Nambu–Goldstone bosons at high density, and use it to derive spectral sum rules for the complex eigenvalues of the Euclidean Dirac operator. In Sect. 3.3 we introduce a non-Hermitian random matrix ensemble and argue that it belongs to the same universality class as the Dirac operator in dense two-color QCD. In Sect. 3.4 the new ChRMT is solved exactly and microscopic spectral densities are obtained in the limit where the matrix size tends to infinity. It is discovered that the microscopic spectrum is quite sensitive to the detuning of mass parameters. In Sect. 3.5 we evaluate the severity of the sign problem in the $\varepsilon$-regime of dense two-color QCD on the basis of the universal correspondence to the new ChRMT. It is observed that, as we detune the masses, the expectation value of the sign of the fermion determinant drops to zero, signaling a severe sign problem. Chapter 4 discusses possible extensions of our analysis on two-color QCD ($\beta = 1$) to other QCD-like theories, namely QCD with isospin chemical potential ($\beta = 2$) and QCD with fermions in a *real* representation of the gauge group ($\beta = 4$) where $\beta$ is the so-called Dyson index. Chapter 5 is devoted to summary and conclusions. In appendices we clarify mathematical conventions of this thesis and present technical details of the calculations that have been omitted in the main text.

## References

1. D.J. Gross, F. Wilczek, Ultraviolet behavior of non-abelian gauge theories. Phys. Rev. Lett. **30**, 1343–1346 (1973)
2. H.D. Politzer, Reliable perturbative results for strong interactions? Phys. Rev. Lett. **30**, 1346–1349 (1973)
3. S. Hashimoto, Spontaneous chiral symmetry breaking on the lattice, PoS **CD09**, 004 (2009).
4. T. Banks, A. Casher, Chiral symmetry breaking in confining theories. Nucl. Phys. **B169**, 103 (1980)
5. JLQCD Collaboration, H. Fukaya et al., Determination of the chiral condensate from 2+1-flavor lattice QCD. Phys. Rev. Lett. 104, 122002 (2010).
6. J.J.M. Verbaarschot, T. Wettig, Random matrix theory and chiral symmetry in QCD. Ann. Rev. Nucl. Part. Sci. **50**, 343–410 (2000)
7. J.B. Kogut, M.A. Stephanov, D. Toublan, On two-color QCD with baryon chemical potential. Phys. Lett. **B464**, 183–191 (1999)
8. J.B. Kogut, M.A. Stephanov, D. Toublan, J.J.M. Verbaarschot, A. Zhitnitsky, QCD-like theories at finite baryon density. Nucl. Phys. **B582**, 477–513 (2000)
9. H. Leutwyler, A.V. Smilga, Spectrum of Dirac operator and role of winding number in QCD. Phys. Rev. **D46**, 5607–5632 (1992)
10. A.V. Smilga, J.J.M. Verbaarschot, Spectral sum rules and finite volume partition function in gauge theories with real and pseudoreal fermions. Phys. Rev. **D51**, 829–837 (1995)
11. M.A. Stephanov, Random matrix model of QCD at finite density and the nature of the quenched limit. Phys. Rev. Lett. **76**, 4472–4475 (1996)
12. J.C. Osborn, Universal results from an alternate random matrix model for QCD with a baryon chemical potential. Phys. Rev. Lett. **93**, 222001 (2004)
13. G. Akemann, J.C. Osborn, K. Splittorff, J.J.M. Verbaarschot, Unquenched QCD Dirac operator spectra at nonzero baryon chemical potential. Nucl. Phys. **B712**, 287–324 (2005)
14. K. Splittorff, J.J.M. Verbaarschot, Phase of the Fermion determinant at nonzero chemical potential. Phys. Rev. Lett. **98**, 031601 (2007)
15. K. Splittorff, J.J.M. Verbaarschot, The QCD sign problem for small chemical potential. Phys. Rev. **D75**, 116003 (2007)
16. J.J.M. Verbaarschot, The spectrum of the QCD Dirac operator and chiral random matrix theory: the three-fold way. Phys. Rev. Lett. **72**, 2531–2533 (1994)
17. B. Vanderheyden, A.D. Jackson, Random matrix models for phase diagrams. Rept. Prog. Phys. **74**, 102001 (2011)
18. T. Kanazawa, T. Wettig, N. Yamamoto, Singular values of the Dirac operator in dense QCD-like theories. JHEP **12**, 007 (2011)

# Chapter 2
# QCD with Chemical Potential and Matrix Models

In this chapter we review QCD at finite temperature and density, QCD with the gauge group SU(2) (also called 'two-color QCD'), and finally chiral random matrix theory for chiral symmetry breaking in QCD. They form the basis of later developments in this thesis.

## 2.1 Phases of QCD

### *2.1.1 Overview*

Quantum Chromodynamics (QCD) at nonzero temperature ($T$), quark chemical potential ($\mu$) and/or external fields (e.g., a magnetic field) is known to exhibit a rich phase structure, which is relevant to various fields of physics including the internal structure of compact stars, early universe and relativistic heavy-ion collisions (see [1–12, 40] for reviews and [13–16] for books). We quote two conjectured QCD phase diagrams in Fig. 2.1. Despite the notable simplicity of the QCD Lagrangian, our understanding on the QCD phase diagram is still limited for various reasons. First of all, QCD is a strongly coupled theory and understanding its non-perturbative behavior theoretically is a hard problem even at zero temperature and density. Even in the region with an asymptotically high density or temperature where the running coupling gets small due to asymptotic freedom, the physics is not literally perturbative: the spatial Wilson loop exhibits an area law even above the deconfinement temperature, while the Fermi surface of quasi-free quarks at high density is unstable toward a formation of Cooper pairs (reviewed in more detail below). Second, experimental observation of quark matter is fairly limited, especially at low temperature and high density. Signatures from compact stars so far have not been sufficient to constrain the QCD phase diagram. Third, the lattice QCD, which has been so successful in gaining insight into QCD at finite temperature as a first-principle formulation of QCD, is afflicted with the so-called sign problem once a nonzero quark chemical potential

T. Kanazawa, *Dirac Spectra in Dense QCD*, Springer Theses,
DOI: 10.1007/978-4-431-54165-3_2, © Springer Japan 2013

is turned on, making the Monte Carlo importance sampling ineffective (see [17–20] for reviews). A general solution to the sign problem, if any, is yet to be discovered, and up to now Monte Carlo simulations of QCD at finite density have been limited to a region of small chemical potential.

A powerful method to look into QCD at finite temperature and density is to use effective models such as the Nambu-Jona-Lasinio model [21–23] as well as its extension [24, 25], which encode chiral symmetry and its spontaneous breakdown correctly. The parameters in the model are a priori unknown and are so tuned that the model reproduces experimental data at $T = \mu = 0$. Although theoretical uncertainty remains regarding the $T$- and $\mu$-dependence of such parameters, we are able to gain various insights into the many body physics of QCD by using those models as a testing ground of theoretical ideas. Other methods include Ginzburg-Landau approaches, linear sigma models, random matrix models, instanton liquid models, strong coupling expansion on the lattice, chiral perturbation theories, the high-density effective theory, and AdS/CFT-inspired models. Nowadays it is even attempted to use ultra-cold atomic gases to mimic dense quark matter [26]. These directions complement each other, covering a wide range of the QCD phase diagram.

In addition to the conventional hadronic phase at low density and temperature and the quark-gluon plasma (QGP) realized at high temperature [27], many phases of QCD have been suggested from various approaches mentioned above, including the neutron superfluid phase, the proton superconducting phase, diverse color-superconducting phases, the meson condensed phase, the gluon condensed phase, the meson current phase, the inhomogeneous chiral symmetry broken phase, and the quarkyonic phase; the superconducting phases may be further classified into the 2SC phase, the Color-Flavor-locked (CFL) phase, the crystalline Fulde-Ferrell-Larkin-Ovchinikov phase, etc. This list will elongate if we incorporate the external magnetic field. (See [28] and references therein.)

In the next section we will review basic features of color superconductivity and the so-called "Color-Flavor-Locked (CFL) Phase" which is believed to be realized in the high density limit of three-flavor QCD. It will be important for our analysis in later chapters.

### 2.1.2 Color Superconductivity and the CFL Phase

The Bardeen-Cooper-Schrieffer (BCS) theory of superconductivity [29–31] is a milestone in the history of theoretical physics [32]. Not only did it explain properties of superconducting metals quantitatively, it also helped people realize the possibility that symmetries respected by the underlying Hamiltonian could be broken spontaneously by the ground state: a first encounter with the concept of Spontaneous Symmetry Breaking (SSB). It led Nambu and Jona-Lasinio to the idea of spontaneous chiral symmetry breaking in QCD [33, 34]. Later the idea of SSB was also exploited in an essential way in the construction of the Standard Model for electroweak interactions [35–37].

The theoretical underpinning of BCS theory is the so-called *Cooper instability*: *the Fermi surface of free electrons becomes unstable once a net attractive interaction is turned on.* For simplicity let us assume $T \simeq 0$ from here on. Suppose adding a pair of electrons on top of the Fermi sphere. The energy increase is given by $2\mu - U$ with $U > 0$ the interaction energy between the two electrons,[1] hence the change in the grand potential reads $\delta\Omega \equiv \delta(E - \mu N) = (2\mu - U) - 2\mu = -U < 0$. The true ground state turns out to be characterized by a condensation of Cooper pairs, consisting of two electrons with opposite momenta on the Fermi surface. The ground state wave function has a definite phase, but an indefinite particle number, thus the U(1) gauge symmetry is broken (or more properly, 'hidden'). Understanding the state of a superconducting metal this way, one can explain many of its pivotal features, such as the Meissner effect and the generation of a *gap* in the electron spectrum.

An interesting point is that *an arbitrarily weak attraction can trigger this qualitative change in the ground state,* in marked contrast to fermionic systems at zero density, where the attraction must exceed a nonzero threshold value to cause the pairing. This difference is essentially due to the fact that the density of states is non-vanishing at zero energy in the presence of the Fermi surface.

Where does the net attraction between electrons come from? The Coulomb interaction is repulsive, even if screened by positive background charges in solids. However the electrons also interact with phonons, arising from the quantization of lattice vibration, and the net interaction is obtained only if the attraction mediated by phonons overcomes the Coulomb interaction. This picture is believed to be correct at least in conventional superconductors, while the origin of attraction is still elusive in exotic high-temperature superconductors.

These ideas for non-relativistic electrons are essentially valid for relativistic fermions as well. In some sense the situation is simpler in QCD than in solids, because the gauge interaction in QCD could be attractive by itself. Consider QCD with asymptotically large quark chemical potential ($\mu \gg \Lambda_{\mathrm{QCD}}$) at $T \simeq 0$. The Fermi surface of quasi-free quarks (due to the small running coupling $g$) would be unstable if there is a net attractive interaction between quarks. As a crude estimate one can look at the one-gluon exchange process. The color structure of a scattering amplitude of two quarks, for general $N_c$, reads

$$(T^A)^{ab}(T^A)^{cd} = -\frac{N_c+1}{4N_c}(\delta^{ab}\delta^{cd} - \delta^{ad}\delta^{bc}) + \frac{N_c-1}{4N_c}(\delta^{ab}\delta^{cd} + \delta^{ad}\delta^{bc}), \quad (2.1)$$

where $\{T^A\}$ ($A = 1, \ldots, N_c^2 - 1$) are the generators of SU($N_c$) in the fundamental representation and $a, b, c, d \in \{1, \ldots, N_c\}$. Therefore attraction occurs in the anti-symmetric channel, namely in the $\bar{\mathbf{3}}$ channel of $\mathbf{3} \otimes \mathbf{3} = \mathbf{6} \oplus \bar{\mathbf{3}}$ for $N_c = 3$. More intuitively, the color flux emanating from the $\bar{\mathbf{3}}$ diquark is obviously smaller than that from two $\mathbf{3}$ quarks, hence a lower energy. The estimate based on one-gluon exchange is not reliable for lower densities, but other studies such as those based on instantons also imply that the $\bar{\mathbf{3}}$ channel is energetically favored. The ground state of dense

[1] We neglect the dependence of $\mu$ on the total number of particles.

quark matter is therefore an aggregate of $\bar{\mathbf{3}}$ diquarks. Since it breaks the SU(3) gauge symmetry spontaneously, this phenomenon is referred to as *color superconductivity* [38, 39].

What about the Lorentz indices of quarks? It is most natural to assume an $s$-wave pairing, since then all quarks near the Fermi surface can contribute to the pairing symmetrically. It leaves us with two possibilities, $\langle qC\gamma_5 q\rangle$ and $\langle qCq\rangle$ for the diquark condensate, corresponding to the positive and negative parity respectively, with $C \equiv i\gamma^2\gamma^0$ the charge conjugation matrix. They are not distinguished by one-gluon exchange interaction, but instanton interaction favors the positive parity [40–42]. The diquark condensate will be dominantly formed in the $J^P = 0^+$ channel even if the instanton effects are quantitatively weak.

The specialty of QCD is that the quarks also have a flavor index, which makes the QCD phase diagram much richer than that of ordinary metals. According to the Pauli exclusion principle the wave function of diquark must be antisymmetric under the exchange of the quarks. Since the color and Lorentz indices are antisymmetric, the flavor indices must also be antisymmetric. The ground state of QCD at large $\mu$ therefore depends on $N_f$ in an essential way.

Color superconductivity leads to (at least) three fundamental consequences:

- (Part of) gluons acquire mass.
- (Part of) quarks acquire mass, namely the BCS gap $\Delta$.
- Chiral symmetry may be broken spontaneously, depending on $N_f$.

The earliest studies on color superconductivity based on perturbative one-gluon exchange gave a value of $\Delta$ that was so small that there seemed to be little chance to observe the effect of color superconductivity in astrophysical scenes. Since then this field of research has remained dormant for more than a decade (but see e.g., [43]), until late 1990s when studies based on non-perturbative instanton interaction reported that a significantly larger $\Delta$ could be realized at modest densities reachable in the interiors of compact stars [40, 44].

There is yet another mechanism which enhances $\Delta$ as compared to the naïve BCS estimate $\Delta \sim \mu e^{-1/g^2}$. In dense medium, color electric forces are subject to Debye screening by quarks, but (static) color magnetic forces are not, to all orders in perturbation theory. Therefore the magnetic forces remain long-ranged. Using a renormalization-group argument around the Fermi surface, Son gave an estimate of $\Delta$ that takes into account the unscreened magnetic interaction [45]

$$\Delta \sim \frac{\mu}{g^5} \exp\left(-\frac{3\pi^2}{\sqrt{2}g}\right) . \tag{2.2}$$

Thus $\Delta$ gets arbitrarily large at asymptotically large $\mu$, even though the interaction will be arbitrarily weak due to asymptotic freedom! It is clear that $\Delta \ll \mu$ at large $\mu$. Equation (2.2) is confirmed by other methods [46–49]. The apparent gauge dependence of (2.2) is expected to disappear when all the sub-leading corrections are properly incorporated. It is exciting that we are able to compute the value of observable from QCD directly, rather than from simplified effective models. At high but not asymptotically high densities, this leading-order expression will not be literally reliable, only serving as an order estimate.

A leading-order estimate on the diquark condensate reads $\langle qq \rangle \sim \mu^2 \Delta / g$. At first sight it seems that a non-vanishing $\langle qq \rangle$ condensate is at odds with the general principle that a local symmetry cannot be spontaneously broken [50]; a local symmetry is not a true symmetry of the Hilbert space but merely reflects a redundancy in our description of the nature, hence no 'breaking'. Therefore a gauge dependent condensate in a particular gauge bears physical relevance only if it is used for the calculation of a gauge-invariant quantity, such as a four-quark condensate $\langle \bar{q}\bar{q}qq \rangle$ and a six-quark condensate $\langle qqqqqq \rangle \sim \langle (uds)(uds) \rangle$ that serves as an order parameter for the $\mathrm{U}(1)_B$ symmetry breaking.

The details of pairing and the symmetry breaking pattern depend on $N_f$. For QCD it is phenomenologically sufficient to consider $N_f \le 3$, i.e., the light quarks $u, d$ and $s$. For $N_f = 2$ ($u$ and $d$) with degenerate masses and densities, the diquark condensate

$$\langle q_a^f q_b^g \rangle \sim C\gamma_5\, \varepsilon_{ab3}\, \varepsilon^{fg} \qquad \text{with } a, b \in \{1, 2, 3\} \text{ and } f, g \in \{1, 2\} \tag{2.3}$$

develops. This is the so-called two-flavor color-superconducting (2SC) phase [40, 44]. The gauge symmetry $\mathrm{SU}(3)_C$ is broken spontaneously to $\mathrm{SU}(2)_C$ and five of the gluons become massive. No global symmetry is broken in this phase: chiral symmetry is obviously intact, while the $\mathrm{U}(1)_B$ symmetry survives as a combination of the original $\mathrm{U}(1)_B$ and $T^8$, the eighth generator of $\mathrm{SU}(3)_C$. As is clear from (2.3) the quarks in the third color do not participate in the pairing. The existence of these gapless modes does not contradict the 't Hooft anomaly matching condition [51]. This phase is relevant at relatively low densities, where $s$ quark does not participate in the dynamics. This picture will be changed to some extent, once the electric charge neutrality and the $\beta$-equilibrium are imposed (see [52] for a review on 2SC).

For $N_f = 3$ various studies suggest that the most favored pairing will be

$$\langle q_a^f q_b^g \rangle \sim C\gamma_5\, \varepsilon^{fgh} \varepsilon_{abh}\, . \tag{2.4}$$

This phase breaks local $\mathrm{SU}(3)_C$ together with the global symmetries in such a way [53]

$$\mathrm{U}(1)_B \times \mathrm{SU}(3)_C \times \mathrm{SU}(3)_R \times \mathrm{SU}(3)_L \to \mathbb{Z}_2 \times \mathrm{SU}(3)_{C+R+L}\, , \tag{2.5}$$

where $\mathrm{SU}(3)_{C+R+L}$ stands for the vectorial subgroup of the three SU(3)'s on the LHS. All gluons are Higgsed and acquire a mass of order $\Delta$ [54–57]. Furthermore the chiral symmetry breaking gives rise to eight Nambu-Goldstone bosons. Although the diquark (2.4) breaks the $\mathrm{U}(1)_{em}$ group for electromagnetism, it survives albeit rotated, by combining with $T^8$ of $\mathrm{SU}(3)_C$. This 'rotated photon' is unscreened in this phase. The intriguing fact is that the right and left chiral symmetries are locked to each other through the color indices of quarks. This phase is termed as the "Color-Flavor-Locked (CFL) phase" [53]. It reminds us of the B phase of the superfluid helium 3, where the spin and the orbital rotations are locked. Precisely speaking, a diquark condensate in the color- and flavor-symmetric channel ($\mathbf{6}_S$, $\mathbf{6}_S$), which is neglected in (2.4), could be nonzero because after SSB (2.5) no symmetry prohibits it from being generated. A perturbative calculation found that the gap in this channel is suppressed by an additional factor of $g$ [58].

The pattern of chiral symmetry breaking in the CFL phase is identical to that of the QCD vacuum; this analogy can be sophisticated further, including the matching of symmetries and spectrums, culminating in the conjecture of "quark-hadron continuity" [59]. It suggests that the CFL phase in the high-density limit of $N_f = 3$ QCD might be continuously connected to the low-density hypernuclear matter. The continuity also takes care of the confinement: since the quarks in the CFL phase have *integer charge* under the rotated electromagnetism and the spectrum contains no massless gluons, it may be regarded as a confining phase! This observation is a non-trivial example of the general wisdom that no order parameter can distinguish the confinement phase from the Higgs phase when the spectrum contains a particle that transforms non-trivially under the center of the gauge group [60, 61]. Whether or not this continuity does really occur on the QCD phase diagram has to be studied by dynamical calculations, not only by symmetries. Despite serious efforts so far, it is still under a debate (see [62–64] for recent works).

Color superconductivity for $N_f > 3$ and $N_f = 1$ were investigated in [58] and in [65, 66], respectively. The former is more or less for an academic interest, whereas the latter is relevant to the dynamics of strange quark at lower densities [67].

The existence of the BCS phase in QCD is firmly established only at low $T$ and high $\mu$. When the temperature is increased, thermal fluctuations tend to destroy the BCS pairing and the phase transition typically occurs at $T_c \approx \Delta$. On the phase diagram, the CFL phase at $T \simeq 0$ will transmute into the QGP phase at high $T$, after experiencing a series of phase transitions [68, 69].

Finally we will comment on chiral anomaly for $\mathrm{U}(1)_A$ and instantons at finite density. As is well known, the flavor-singlet axial current has a non-vanishing divergence at $\mu = 0$,

$$\partial_\mu J_5^\mu = -\frac{N_f g^2}{32\pi^2}\varepsilon^{\alpha\beta\gamma\delta}F^A_{\alpha\beta}F^A_{\gamma\delta}. \tag{2.6}$$

This relation can be extended to $\mu \neq 0$ through an explicit calculation of the triangle diagram [70]. Zero-mode solutions at $\mu \neq 0$ were constructed explicitly in [71]. Does it mean that the chiral anomaly is independent of $\mu$? The answer is No. Since this relation holds at the operator level and has nothing to do with the physical states for which expectation values of observables are calculated, we must do dynamical calculations to see the behavior of anomaly and instantons at $\mu \neq 0$.

Physically one can expect that the electric fields inside instantons are subject to Debye screening due to condensed quarks. Another simple reason for the suppression of instantons is that the contribution of a instanton comes with a factor $\mathrm{e}^{-1/g^2}$, which gets arbitrarily small in the high-density limit due to asymptotic freedom. According to detailed perturbative calculations [72], the instanton density for size $\rho$ at large $\mu$ is given by

$$n(\rho, \mu) \sim n(\rho, 0)\mathrm{e}^{-N_f \mu^2 \rho^2} . \tag{2.7}$$

Therefore the infrared problem associated with large instantons at $\mu = 0$ is removed by the screening factor $\mathrm{e}^{-N_f \mu^2 \rho^2}$, making semi-classical calculations at large $\mu$ tractable. It is misleading, however, to say based on (2.7) that the instanton effects are suppressed *exponentially* at large $\mu$. In calculating physical observables one typically faces an integral of the form $\int_0^\infty d\rho \ \rho^{\#} n(\rho, \mu) \sim \int_0^\infty d\rho \ \rho^{\#'} \mathrm{e}^{-8\pi^2/g(\rho)^2} \mathrm{e}^{-N_f \mu^2 \rho^2}$. Once the $\rho$-integration is done, with the running of $g(\rho)$ taken into account, one finds a negative power of $\mu$, not an exponential. Important instanton-induced observables, such as the $\eta'$ mass and the chiral condensate in the CFL phase, are found to be suppressed at large $\mu$ by inverse powers of $\mu$; see [73–76] for more details. It is worth stressing that both the asymptotic freedom and the Debye screening are essential in these calculations.

The estimate (2.7) is not trustworthy at moderate densities. The $\mu$-dependence of anomaly-induced interaction at low $\mu$ is difficult to compute theoretically, but can nonetheless have a qualitative impact on the QCD phase diagram [25, 77, 78].

To gain insight into the QCD ground state at low $\mu$ one often uses effective models of QCD, such as the NJL model, but they are prone to several problems. First, finding the ground state of a many-body system is a hard problem even in the NJL model; one picks up a particular state, plug it into the model to calculate its ground state energy, and check if it is lower than those of the other states tested before. By this procedure one can never guarantee that the tested state is more stable than *any other possible state*. Secondly, the determination of the phase boundaries is a quantitative problem in its nature; if there are many local minima in the grand potential (corresponding to different symmetry breaking patterns), we face a genuinely *quantitative* problem of determining which minimum has the lowest energy. Symmetry principle is not helpful in this regard. Any uncertainties in the effective model, e.g., the $(T, \mu)$-dependence of coefficients, may potentially lead to a sizable change in the phase structure. This is a dangerous situation. It seems that a fundamentally new approach is needed for further developments in this field.

We would like to summarize Sect. 2.1. QCD at nonzero temperature and density is full of interesting phases. The global as well as local symmetries of QCD are either spontaneously or explicitly broken in various ways, which makes the determination of phase boundaries and critical points a highly challenging problem. Only at sufficiently high density do we have a solid understanding of called quark matter: a color superconducting state, called the Color-Flavor-Locked (CFL) phase. At moderate densities the coupling grows and a microscopic approach to QCD is not viable in general, except on the $\mu = 0$ axis where lattice QCD simulation does not suffer from the sign problem. Theoretical understanding of the intermediate density region between the hadronic phase and the color-superconducting phase remains an unsolved challenging problem.

## 2.2 Two-Color QCD at Low Density

### 2.2.1 Basics

In this section we review basic features of $N_c = 2$ QCD at finite density and explain the motivation to investigate this theory. (The focus of discussions below is on the low-density region. For asymptotically large density we will have different physics, see Chap. 3 Sect. 3.2.)

As a means of sidestepping the severe sign problem in lattice QCD at finite density, several QCD-like theories have been investigated numerically and theoretically. The primary advantage of studying them is that they are free from the sign problem even at finite density (for at least a particular choice of parameters in the theory) so that *a direct comparison between numerical results and analytical predictions is feasible.* Their list includes QCD with imaginary chemical potential [79], QCD with isospin chemical potential [80, 81] and QCD with quarks in a real or pseudoreal representation of the gauge group [82].[2] (The NJL model is also simulable at nonzero density [85, 86].) While being interesting on their own, these theories also serve as theoretical laboratories for new ideas and methods developed for the dense quark matter; they allow us to check predictions of effective models by comparison with lattice data, test new simulation algorithms for lattice QCD, etc.

In particular, QCD for gauge group SU(2) with quarks in the fundamental representation (**2**), which is commonly called "two-color QCD", is one of the theories with pseudoreal fermions mentioned above. It has been intensively studied by various methods:

- Lattice simulations [87–107],
- The NJL model [108–113] and the PNJL model [114],
- Chiral perturbation theories [82, 115–124],
- Dressed-rainbow gap equation [125]
- The instanton liquid model [71, 126],
- Instanton-based calculations at high density [44, 74, 75],
- Linear sigma models [127, 128],
- The strong-coupling limit on the lattice [129–136],
- Random matrix models [137–140].

Before delving into technical details, let us list unique properties of two-color QCD which distinguish it from ordinary three-color QCD:

---

[2] [Mathematical footnote] In representation theory, a representation is called *real* if the entries of the matrices expressing the group elements are real. A representation is called *pseudoreal* if it is isomorphic to its complex conjugate but it cannot be expressed as real matrices [83, 84]. Examples of pseudoreal representation include **2** of SU(2), **20** of SU(6), the fundamental representation of Sp($2n$), and the spinor representation of Spin($8n-3$), Spin($8n-4$) and Spin($8n-5$) for $n \in \mathbb{N}$. (Spin(n) is the universal cover of SO($n$).)

(1) All color-singlet excitations are bosons.
(2) Spontaneous breaking of flavor symmetry occurs, but in an exotic way.
(3) Some of the pions are charged under the baryon number (= 'baryonic pion').
(4) At high density it is not a color superconductor but a superfluid.
(5) Sign problem is absent at nonzero $\mu$ when $N_f$ is even and the quark masses are pairwise degenerate.

These properties are not mutually independent but are more or less related to each other at a deeper level as we review below. The basic group-theoretical fact is that the fundamental representation **2** of SU(2) is isomorphic to its complex conjugate, $\bar{\mathbf{2}}$, as seen from

$$\sigma_2(-\sigma_a^*)\sigma_2 = \sigma_a\,, \qquad a = 1, 2, 3. \tag{2.8}$$

Using (2.8) as well as formulas in the Appendix A.1, one can prove the anti-unitary symmetry of the Euclidean Dirac operator $\mathcal{D}(\mu) \equiv \gamma_\mu \mathcal{D}_\mu - \mu\gamma_0$:

$$\mathcal{D}(\mu)^* = P^{-1}\mathcal{D}(\mu)P \qquad \text{for } P \equiv \gamma_5 C\tau_2\,, \tag{2.9}$$

where $\tau_2$ is the color generator and $D_\mu \equiv \partial_\mu + igA_\mu^a\tau_a$. Using this symmetry we obtain

$$\det[\mathcal{D}(\mu)+m]^* = \det[P^{-1}\{\mathcal{D}(\mu)+m\}P] = \det[\mathcal{D}(\mu)+m] \tag{2.10}$$

$$\therefore\ \det[\mathcal{D}(\mu)+m] \in \mathbb{R} \tag{2.11}$$

The positivity of the measure is ensured from (2.10) when $N_f$ is even and the masses are pairwise degenerate. This is why a direct lattice simulation is feasible for two-color QCD even at nonzero density.

Using (A.18) in the Appendix A.1 one finds

$$\sum_{f=1}^{N_f} \bar{\psi}_f \gamma_\mu D_\mu \psi_f = \sum_{f=1}^{N_f} \begin{pmatrix} \psi_R^* \\ \psi_L^* \end{pmatrix}_f^T \begin{pmatrix} \mathbf{0} & \mathbf{1} \\ \mathbf{1} & \mathbf{0} \end{pmatrix} \begin{pmatrix} \mathbf{0} & -i\sigma_\mu D_\mu \\ i\sigma_\mu^\dagger D_\mu & \mathbf{0} \end{pmatrix} \begin{pmatrix} \psi_R \\ \psi_L \end{pmatrix}_f \tag{2.12}$$

$$= i\sum_{f=1}^{N_f} \begin{pmatrix} \psi_R^* \\ \psi_L^* \end{pmatrix}_f^T \begin{pmatrix} \sigma_\mu^\dagger D_\mu & \mathbf{0} \\ \mathbf{0} & -\sigma_\mu D_\mu \end{pmatrix} \begin{pmatrix} \psi_R \\ \psi_L \end{pmatrix}_f, \tag{2.13}$$

with $\sigma_\mu \equiv (i\mathbf{1}, \sigma_1, \sigma_2, \sigma_3)$. ($\sigma_a$ and $\tau_a$ act on the spinor and the color indices, respectively.) Equation (2.12) is clearly invariant under $\mathrm{U}(N_f)_R \times \mathrm{U}(N_f)_L$. Defining $\tilde{\psi}_R \equiv \sigma_2\tau_2\psi_R^*$ and using (2.8) one finds

**Table 2.1** The symmetries of two-color QCD [123]. In the first two lines, $\mathrm{U}(1)_B$ is contained in $\mathrm{SU}(2N_f)$ and $\mathrm{Sp}(2N_f)$, respectively. $(\mathbb{Z}_{2N_f})A$ is the anomaly-free subgroup of the axial $\mathrm{U}(1)_A$ symmetry. In the last two lines, even $N_f$ is assumed

| | | | |
|---|---|---|---|
| $\mu = 0$ | $\langle\bar{\psi}\psi\rangle = 0$ | $\langle\psi\psi\rangle = 0$ | $\mathrm{SU}(2N_f)$ $[\supset (\mathbb{Z}_{2N_f})_A]$ |
| | $\langle\bar{\psi}\psi\rangle \neq 0$ | | $\mathrm{Sp}(2N_f)$ $[\supset (\mathbb{Z}_2)_A]$ |
| $\mu \neq 0$ | $\langle\bar{\psi}\psi\rangle = 0$ | $\langle\psi\psi\rangle = 0$ | $\mathrm{SU}(N_f)_L \times \mathrm{SU}(N_f)_R \times \mathrm{U}(1)_B$ $[\supset (\mathbb{Z}_{2N_f})_A]$ |
| | $\langle\bar{\psi}\psi\rangle \neq 0$ | | $\mathrm{SU}(N_f)_V \times \mathrm{U}(1)_B$ $[\supset (\mathbb{Z}_2)_A]$ |
| $\mu \neq 0$ | $\langle\bar{\psi}\psi\rangle = 0$ | $\langle\psi\psi\rangle \neq 0$ | $\mathrm{Sp}(N_f)_L \times \mathrm{Sp}(N_f)_R$ $[\supset (\mathbb{Z}_2)_L \times (\mathbb{Z}_2)_R]$ |
| | $\langle\bar{\psi}\psi\rangle \neq 0$ | | $\mathrm{Sp}(N_f)_V$ $[\supset (\mathbb{Z}_2)_B]$ |

$$\sum_{f=1}^{N_f} \bar{\psi}_f \gamma_\mu D_\mu \psi_f \cong i \sum_{f=1}^{N_f} \begin{pmatrix} \tilde{\psi}_R^* \\ \psi_L^* \end{pmatrix}_f^T \begin{pmatrix} -\sigma_\mu D_\mu & \mathbf{0} \\ \mathbf{0} & -\sigma_\mu D_\mu \end{pmatrix} \begin{pmatrix} \tilde{\psi}_R \\ \psi_L \end{pmatrix}_f \tag{2.14}$$

$$= -i \sum_{F=1}^{2N_f} \Psi_F^\dagger \sigma_\mu D_\mu \Psi_F\,, \qquad \text{for } \Psi_F \equiv \begin{pmatrix} \tilde{\psi}_R \\ \psi_L \end{pmatrix}_f, \tag{2.15}$$

where $\cong$ stands for an equality up to a total derivative. Therefore the action is actually invariant under the larger flavor group $\mathrm{U}(2N_f) = \mathrm{U}(1)_A \times \mathrm{SU}(2N_f)$. Note that $\mathrm{U}(1)_B \subset \mathrm{SU}(2N_f)$. This special symmetry of *massless* quarks in two-color QCD is called the Pauli-Gürsey symmetry [141, 142]. This extended flavor symmetry also shows up in QCD with quarks in an arbitrary pseudoreal or real representation of the gauge group, and two-color QCD is just a special case.

The flavor-diagonal mass term can be written as

$$m \sum_{f=1}^{N_f} \bar{\psi}_f \psi_f = -\frac{1}{2} m \sum_{F,G=1}^{2N_f} \Psi_F^T \sigma_2 \tau_2 (I_{2N_f})^{FG} \Psi_G + \text{h.c.}, \tag{2.16}$$

where we introduced a $2N_f \times 2N_f$ antisymmetric flavor matrix

$$I_{2N_f} \equiv \begin{pmatrix} \mathbf{0} & -\mathbf{1}_{N_f} \\ \mathbf{1}_{N_f} & \mathbf{0} \end{pmatrix}. \tag{2.17}$$

Equation (2.16) is invariant only under $\mathrm{Sp}(2N_f) \subset \mathrm{SU}(2N_f)$.[3] The symmetries of two-color QCD are summarized in Table 2.1.

There is yet another possible external source term. Since the diquarks in two-color QCD are color singlet (unlike in QCD), one can add a diquark source of the form

[3] The symplectic group is defined as $\mathrm{Sp}(2n) = \{g \in \mathrm{U}(2n) \mid g^T I_{2n} g = I_{2n}\}$. We adopt a convention in which $\mathrm{Sp}(2) \cong \mathrm{SU}(2)$.

$$-\frac{j}{2}\sum_{f,g=1}^{N_f}\psi_f^T C\gamma_5 i\tau_2 (I_{N_f})^{fg}\psi_g + \text{h.c.} = \frac{j}{2}\sum_{F,G=1}^{2N_f}\Psi_F^T\sigma_2\tau_2\begin{pmatrix} iI_{N_f} & \mathbf{0} \\ \mathbf{0} & iI_{N_f}\end{pmatrix}_{FG}\Psi_G, \tag{2.18}$$

where $C$ is the charge conjugation matrix and $\tau_2$ is a color generator. In (2.18) it is tacitly assumed that $N_f$ is *even*. ($I_n$ is defined only for even $n$.) The case of odd $N_f$ (not treated in [82, 116]) calls for a separate analysis; due to the Pauli principle we must have an antisymmetric flavor matrix, and any antisymmetric matrix of odd dimension has (at least) one zero eigenvalue. Thus one of the flavors disappears from the diquark pairing (2.18) and makes the symmetry breaking more complicated. We henceforth assume that $N_f$ is even.

Such a source is useful to investigate the diquark condensation in lattice simulations. One can show that (2.18) breaks the flavor symmetry explicitly as $\mathrm{SU}(2N_f) \to \mathrm{Sp}(2N_f)$. This is the same as what the mass term (2.16) does. Indeed, the mass term $\Psi^T I_{2N_f}\Psi$ and the diquark source $\Psi^T\begin{pmatrix} I_{N_f} & \mathbf{0} \\ \mathbf{0} & I_{N_f}\end{pmatrix}\Psi$ transform into each other under the action of $\mathrm{SU}(2N_f)$. Therefore adding a mass term alone is equivalent to adding a diquark source alone, except that the diquark source breaks the exact $\mathrm{U}(1)_B$ symmetry, while the mass term breaks the anomalous $\mathrm{U}(1)_A$ symmetry.

It is important to note that the equivalence does not hold at $\mu \neq 0$ because the $\mathrm{SU}(2N_f)$ symmetry is explicitly broken by $\mu$ (cf. Table 2.1). In [143] it was shown that the diquark source is a useful probe for the singular value spectrum of the Dirac operator at $\mu \neq 0$. Although this is an intriguing discovery, we will be mainly concerned with the case of vanishing diquark source in the remainder of this thesis for the simplicity of exposition.

It was argued in [144, 145] that, in the chiral limit, the chiral condensate $\langle\sum_{f=1}^{N_f}\bar{\psi}_f\psi_f\rangle$ would develop and break the symmetry spontaneously as[4]

$$\mathrm{SU}(2N_f) \to \mathrm{Sp}(2N_f). \tag{2.19}$$

This claim was 'proven' in [146]; the argument goes as follows. Using a Vafa-Witten type arguments [147] one can show that $\mathrm{Sp}(2N_f)$ is not spontaneously broken. Since $\mathrm{Sp}(2N_f)$ is a maximal subgroup of $\mathrm{SU}(2N_f)$, the symmetry must be either broken to $\mathrm{Sp}(2N_f)$ or be unbroken at all. Then 't Hooft anomaly matching condition comes into play. Since color confinement in two-color QCD forbids color-singlet composite fermions, one cannot use them to produce $\mathrm{SU}(2N_f)$ global anomalies of elementary quarks. Thus the symmetry must be spontaneously broken.[5]

This formal proof is suggestive but not rigorous because of serious loopholes in the original Vafa-Witten argument [148–150]; operators composed of fermion bilinears invalidate the proof. However the symmetry breaking pattern (2.19) has

[4] They take the limit $m \to +0$ with $j \equiv 0$. If another possible limit $j \to +0$ with $m \equiv 0$ is taken, the diquark condensate will be formed, rather than the chiral condensate.

[5] For *real* quarks a special care is needed [146], as gluons may screen quarks to form color-singlet massless fermions.

**Table 2.2** The 'three-fold way' of chiral symmetry breaking in QCD with $N_f$ Dirac fermions in various representations [144, 151, 152]

| Color representation of quarks | Symmetry breaking pattern | Examples |
|---|---|---|
| Pseudoreal | $\mathrm{SU}(2N_f) \to \mathrm{Sp}(2N_f)$ | Fundamental rep. of SU(2) |
| Complex | $\mathrm{SU}(N_f)_R \times \mathrm{SU}(N_f)_L \to \mathrm{SU}(N_f)_V$ | Fundamental rep. of $\mathrm{SU}(N > 2)$ |
| Real | $\mathrm{SU}(2N_f) \to \mathrm{SO}(2N_f)$ | Adjoint rep. of $\mathrm{SU}(N)$ |

been checked so far by numerous direct lattice simulations, putting (2.19) on a firm ground.

In Table 2.2 we summarize the patterns of spontaneous chiral symmetry breaking for three classes of fermions in 4-dimensional gauge theories.

### 2.2.2 Why ChPT is Worth Doing

Here and in the next section we would like to give a brief review of the studies of two-color QCD based on the chiral perturbation theory (ChPT) [82, 116]. The reason is twofold: first, it is solely based on the symmetries of the microscopic Lagrangian and gives a *model-independent* description of the low-energy dynamics. Second, it constitutes important preliminary knowledge for my original work presented in a later chapter of this thesis.

First of all, what is ChPT? Since there exists vast literature on this topic (see e.g., [153–156] for reviews and [157] for a book), We would only summarize important features. The chiral symmetry breaking (ChSB) $\mathrm{SU}(N_f)_R \times \mathrm{SU}(N_f)_L \to \mathrm{SU}(N_f)_V$ in three-color QCD in the chiral limit (for not too large $N_f$) at zero temperature and baryon density has been established firmly by experiments and lattice simulations [158]. Regarding this, there are (at least) three questions to ask:

1. Why does ChSB occur?
2. How does ChSB occur?
3. What can we predict based on the fact of ChSB?

In a nutshell, ChPT is a framework that addresses the final question. Currently we do not yet know the answer to the first question; ChSB is not proven from first principles even today.[6] While the origin of ChSB is an important and still active area of research, we will not tackle this issue in the rest of this thesis. Regarding the second question, we will review the approach called Chiral Random Matrix Theory in the next section.

ChPT is a kind of Effective Field Theory (see [161–165] for reviews). The idea of effective theory method is to simplify the theory by integrating out all degrees of freedom except those that are relevant in the energy scale of interest. This strategy

[6] There are special cases where ChSB is proven with mild assumptions; see [159, 160].

works especially well when there is a hierarchy in the spectrum, and it typically happens in theories with spontaneous breaking of global symmetries, where the Nambu-Goldstone (NG) bosons are far lighter than other massive particles. In rare cases the procedure of integrating-out can be done explicitly (e.g., the Euler-Heisenberg Lagrangian), but it is technically too involved in general. Then we have to make maximal use of symmetries to constrain the form of the effective theory. In three-color QCD, the low-energy effective Lagrangian of pions based on a nonlinear realization of chiral symmetry reads[7]

$$\mathcal{L} = \frac{F_\pi^2}{4}\mathrm{tr}[\partial_\mu U^\dagger \partial^\mu U] + \Sigma_0 \,\mathrm{Re}\,\mathrm{tr}[MU] + \dots, \quad U \in \mathrm{SU}(3), \; M = \mathrm{diag}(m_u, m_d, m_s), \tag{2.20}$$

where the dots stand for other terms allowed by symmetries. Even without explicit calculations one can learn about the dynamics of pions from (2.20); e.g., owing to the absence of non-derivative terms at $M = 0$ the interactions between pions must become arbitrarily weak in the low-energy limit. The low energy theorems of this kind have played a quite important role as verifications of our understanding of pions as the NG bosons of chiral symmetry breaking. Later it was even extended to incorporate massive particles such as baryons and vector mesons, achieving reasonable agreement with experiments.

To sort out infinitely many possible terms, one uses the Weinberg's power counting rule [166, 167], according to which

$$\partial_\mu \sim \frac{1}{L} \sim \mathcal{O}(p), \qquad m \sim \mathcal{O}(p^2), \tag{2.21}$$

and align terms according to the power of $p$. ($L$ denotes the linear size of the box. The momentum is discretized in units of $1/L$.) This is an explicit meaning of saying that ChPT is a low-energy expansion in terms of external momenta and small external fields (masses, temperature, etc.). Since higher-order vertices in the number of pions are suppressed by inverse powers of $\Lambda_{\mathrm{QCD}}$, or more precisely $4\pi F_\pi$(~1 GeV), the dimensionless expansion parameter of ChPT is $p/4\pi F_\pi$.

It is obvious that ChPT is perturbatively non-renormalizable. Nevertheless ChPT is renormalizable *at any fixed order* of the $p$-expansion. The divergences from one-loop diagrams of $\mathcal{O}(p^2)$ vertices are absorbed by tree diagrams of $\mathcal{O}(p^4)$ vertices, and the divergences from two-loop diagrams of $\mathcal{O}(p^2)$ vertices are absorbed by one-loop diagrams of $\mathcal{O}(p^4)$ vertices and tree diagram of $\mathcal{O}(p^6)$ vertices, etc.

One might wonder why on earth the effective Lagrangian (2.20), containing infinitely many low-energy constants such as $f_\pi$ and $\Sigma_0$, could have a predictive power. The answer is that this expansion must be terminated at a specific order of $p$ which is high enough to calculate the observable of interest with the desired accuracy. The number of necessary low-energy constants is no longer infinite, thus one can fix them by comparison with experimental data. Then one can use them to compute infinitely many kinds of scattering amplitude; the theory indeed has a predictive power, *up to a specified accuracy.*

---

[7] We neglect electromagnetic interactions and the Wess-Zumino-Witten term.

Above the intrinsic cutoff scale $\Lambda_{\text{QCD}}$, ChPT is no longer effective, because

- Particles other than pions (such as vector mesons) begin to participate in the dynamics, or alternatively,
- At $p \gtrsim \Lambda_{\text{QCD}}$ or $M \gtrsim \Lambda_{\text{QCD}}$ the contributions from higher order terms are not suppressed, thus invalidating the perturbative expansion itself.

One might wonder what if we use a *renormalizable* linear $\sigma$ model instead of the non-renormalizable chiral Lagrangian (2.20). Actually the chiral symmetry warrants that their predictions are indistinguishable at the leading order, both governed by the low-energy constants $F_\pi$ and $\Sigma_0$. The difference emerges if sub-leading contributions are taken into account. To see this, note that in the linear $\sigma$ model, the renormalizability guarantees that only finite number of parameters are involved, whereas more and more parameters show up at higher orders in the chiral Lagrangian. The linear $\sigma$ model then gives rise to nontrivial constraints between low-energy constants which are not fulfilled by the observed values; it is too restrictive [168].

### 2.2.3 ChPT for Two-Color QCD at Small $\mu$

Let us now turn to the application of ChPT to two-color QCD. The spontaneous flavor symmetry breaking (2.19) by nonzero $\langle\bar{\psi}\psi\rangle$ produces NG bosons. Their number is

$$\dim \mathrm{SU}(2N_f) - \dim \mathrm{Sp}(2N_f) = (2N_f)^2 - 1 - 2N_f^2 - N_f \tag{2.22}$$

$$= 2N_f^2 - N_f - 1. \tag{2.23}$$

The coset $\mathrm{SU}(2N_f)/\mathrm{Sp}(2N_f)$ is parametrized by

$$\Sigma = U I_{2N_f} U^T \qquad \text{with} \qquad U \in \mathrm{SU}(2N_f)\,, \tag{2.24}$$

where the antisymmetric flavor matrix $I_{2N_f}$ is defined in (2.17). Since $\Sigma$ is manifestly invariant under the change of variables $U \to UY$ for ${}^\forall Y \in \mathrm{Sp}(2N_f)$, $\Sigma$ indeed parametrizes the degrees of freedom in the coset $\mathrm{SU}(2N_f)/\mathrm{Sp}(2N_f)$. Based on symmetry arguments, one can write down the leading-order effective Lagrangian at $\mu = 0$, with both the mass term and the diquark source, as

$$\mathcal{L} = \frac{F^2}{2}\mathrm{tr}[\partial_\mu \Sigma^\dagger \partial_\mu \Sigma] + mG\,\mathrm{Re}\,\mathrm{tr}[I_{2N_f}\Sigma] + jG\,\mathrm{Re}\,\mathrm{tr}\left[\begin{pmatrix} iI_{N_f} & \mathbf{0} \\ \mathbf{0} & iI_{N_f} \end{pmatrix}\Sigma\right]. \tag{2.25}$$

The Gell-Mann–Oaks–Renner relation is obtained by expanding $\Sigma$ around $I_{2N_f}$ to second order as

$$m_\pi^2 = \frac{\sqrt{m^2 + j^2}\,G}{F^2}\,, \tag{2.26}$$

where $G \equiv \dfrac{|\langle\bar{\psi}\psi\rangle_0|}{2N_f}$ with $\langle\bar{\psi}\psi\rangle_0$ the chiral condensate at $m = \mu = 0$.

In three-color QCD, ChPT does not explicitly depend on $\mu$, since the pions $\sim\bar{\psi}\gamma_5\psi$ carry no baryon charge. The partition function at $T = 0$ is independent of $\mu$, up to a critical $\mu_c$ above which a nonzero baryon density develops. Here $\mu_c$ is order of one-third of the baryon mass, $\sim$300 MeV.

In two-color QCD, the situation is quite different. The Lagrangian at $j = m = 0$ reads

$$\sum_{f=1}^{N_f} \overline{\psi}_f(\gamma_\mu D_\mu - \mu\gamma_0)\psi_f = -i\Psi^\dagger(\sigma_\mu D_\mu + i\mu\mathcal{B})\Psi \tag{2.27}$$

$$\equiv -i\Psi^\dagger\sigma_\mu(D_\mu + \mu B_\mu)\Psi\,, \tag{2.28}$$

where we introduced a $2N_f \times 2N_f$ baryon-charge matrix

$$\mathcal{B} \equiv \begin{pmatrix} \mathbf{1}_{N_f} & \mathbf{0} \\ \mathbf{0} & -\mathbf{1}_{N_f} \end{pmatrix}, \tag{2.29}$$

and a four-vector $B_\mu \equiv (\mathcal{B}, 0, 0, 0)$. Therefore the quark chemical potential explicitly breaks the flavor symmetry down to the usual one:

$$\mathrm{SU}(2N_f) \to \mathrm{SU}(N_f)_R \times \mathrm{SU}(N_f)_L\,. \tag{2.30}$$

It indicates that $\mu$ appears in ChPT as an explicit symmetry breaking field. To see this, we shall use the fact that (2.27) is actually invariant under a *local* $\mathrm{SU}(2N_f)$ transformation [116] under which

$$\Psi \to V\Psi, \qquad B_\mu \to VB_\mu V^\dagger + \frac{1}{\mu}V\partial_\mu V^\dagger \qquad \text{for } V \in \mathrm{SU}(2N_f)\,. \tag{2.31}$$

This local symmetry should carry over to the effective theory. For simplicity, set $j \equiv 0$. Applying this requirement to (2.25), one gets

$$\mathcal{L} = \frac{F^2}{2}\left(\mathrm{tr}[\nabla_\mu\Sigma^\dagger\nabla_\mu\Sigma] + 2m_\pi^2 \mathrm{Re}\,\mathrm{tr}[I_{2N_f}\Sigma]\right), \tag{2.32}$$

$$\nabla_\mu\Sigma \equiv \partial_\mu\Sigma + \mu(B_\mu\Sigma + \Sigma B_\mu^T). \tag{2.33}$$

Surprisingly, the coefficients of $\mu$-dependent terms are uniquely fixed by the local symmetry alone. No new low-energy constant appears up to the second order in $\mu$.

**Table 2.3** Mean field predictions for condensates in two-color QCD

| Phase | $\langle\bar{\psi}\psi\rangle$ | $\langle\psi\psi\rangle$ |
|---|---|---|
| $\mu < m_\pi/2$ | $\langle\bar{\psi}\psi\rangle_0$ | 0 |
| $\mu > m_\pi/2$ | $\langle\bar{\psi}\psi\rangle_0\left(\dfrac{m_\pi}{2\mu}\right)^2$ | $\langle\bar{\psi}\psi\rangle_0\sqrt{1-\left(\dfrac{m_\pi}{2\mu}\right)^4}$ |

This expression can be regarded as the leading-order terms within a counting scheme [119, 120]

$$\partial_\mu \sim \mu \sim \mathcal{O}(p) \quad \text{and} \quad m \sim j \sim \mathcal{O}(p^2)\,. \tag{2.34}$$

The mean field approximation to (2.32) (i.e., without kinetic terms) reveals that a second-order superfluid phase transition occurs at $\mu_c \equiv m_\pi/2$. For $0 \le \mu < m_\pi/2$ the Lagrangian $\mathcal{L}$ takes minimum at $\Sigma = I_{2N_f}$, while for $\mu \ge m_\pi/2$ the minimum moves away from $I_{2N_f}$, as a function of $\mu$. The order of the transition and the value of $\mu_c$ remain unchanged even at next-to-leading order [119]. This transition is caused by the Bose-condensation of baryonic pions. The quantum critical point $\mu_c$ moves to zero in the chiral limit: the existence of a light particle charged under $\mathrm{U}(1)_B$ brings the superfluid transition down to a much lower $\mu$!

This superfluid phase transition has been verified in a number of effective models, such as the (P)NJL models, strong-coupling expansion on the lattice and random matrix models. It occurs universally in any models that incorporate the flavor symmetry (or the Pauli-Gürsey symmetry) as well as its breaking correctly.

By differentiating the vacuum energy by $m$ or $j$ one can derive the $\mu$-dependence of the chiral condensate and the diquark condensate. The mean field result [82] is shown in Table 2.3.

An interesting feature here is that $\langle\bar{\psi}\psi\rangle$ goes to zero then the chiral limit is taken at fixed $\mu > 0$, in contrast to three-color QCD where it tends to a nonzero value. Note also that the square sum $\langle\bar{\psi}\psi\rangle^2 + \langle\psi\psi\rangle^2$ does not depend on $\mu$; the chiral condensate 'rotates' into the diquark condensate.

The inclusion of $j \neq 0$ will render the superfluid transition a crossover. Usually $j \neq 0$ is employed in lattice simulations to suppress infrared fluctuations in the superfluid phase and to make the algorithm ergodic.

The analysis was extended to finite temperature [120] via the 1-loop contribution of pions. It was found that the superfluid transition is second order for lower $T$ and *first order* for higher $T$, in units of $m_\pi$. Therefore there must be a tricritical point on the phase diagram. It was also seen in lattice simulations. Figure 2.2 shows a schematic phase diagram from [96].

However, this tricritical point has never been found in past calculations based on chiral effective models. This apparent failure of model calculations seems to be due to the fact that, as emphasized in [111], the mean-field approximation for chiral models only includes quasifermionic modes but fails to include thermal NG excitations. To improve this, we will have to go beyond the mean-field approximation, see [113] for an analytical attempt in this direction for the NJL model.

The *parameter-free* predictions in Table 2.3 have been tested by many lattice simulations, and the agreement is generally good at small $\mu$ (see e.g., [96]). The $\mu$-independence of $\langle\bar{\psi}\psi\rangle^2 + \langle\psi\psi\rangle^2$ was also verified to a good accuracy. At larger $\mu$, however, the agreement is less impressive, because the subleading contributions in ChPT are no longer negligible. Actually $\langle\psi\psi\rangle$ continues rising beyond $\langle\bar{\psi}\psi\rangle_0$, instead of converging to it, at higher $\mu$.

Even without the sign problem it is a nontrivial challenge to investigate the high density region using lattice QCD: when $\mu$ becomes comparable to the cutoff scale $\sim 1/a$ ($a$: lattice spacing), the discrete nature of the lattice shows up. As the quark number density per site approaches the maximum value $\sim N_f N_c$ allowed by the Pauli principle, quarks are essentially freezed, and the theory resembles pure Yang-Mills theory. To overcome this we need to achieve both a large volume and a small lattice spacing.

We finally note that the symmetry breaking patterns are modified for staggered fermions [90]. The $N$ flavors of staggered fermions in the fundamental (**2**) or adjoint (**3**) representation of SU(2) exhibit following patterns of spontaneous chiral symmetry breaking by $\langle\bar{q}q\rangle$:

$$\text{fundamental: } \mathrm{U}(2N) \to \mathrm{O}(2N) \qquad \text{adjoint: } \mathrm{U}(2N) \to \mathrm{Sp}(2N)\,. \tag{2.35}$$

Therefore the symmetries of fundamental and adjoint fermions are *reversed*. This peculiar fact has been known for long [169–171]. It is unclear if the correct pattern of symmetry breaking will be restored in the continuum limit. We are aware of no study confirming this restoration.

## 2.3 Chiral Random Matrix Theory

### *2.3.1 Dirac Spectrum and ChPT*

In this section we review past studies on the Dirac spectrum in three-color QCD at $T = \mu = 0$. As is well known, the current quark mass $m_f$ acts as an external source for ChSB. It helps the theory pick up one of the degenerate ground states related by global chiral transformations. The fact that even *infinitesimal* $m_f$ can have finite impact on the macroscopic scale, combined with the representation of the partition function in terms of Dirac operator eigenvalues $\{i\lambda_n\}_n$,[8]

$$Z(\{m_f\}) \equiv \int [dA_\mu] \int [d\bar{\psi}][d\psi]\, e^{-\int d^4x \left\{\sum_{f=1}^{N_f} \bar{\psi}_f (D+m_f)\psi_f + \frac{1}{2}\mathrm{tr}[FF]\right\}} \tag{2.36}$$

[8] Here we assume that the system is in a finite Euclidean torus of volume $V_4 = L^3/T \simeq L^4$ with a proper UV cutoff. In our convention the eigenvalues of the Euclidean Dirac operator are purely imaginary ($^\forall\lambda_n \in \mathbb{R}$).

$$= \int [dA_\mu] \prod_{f=1}^{N_f} \prod_n (i\lambda_n + m_f)\, e^{-\frac{1}{2}\int d^4x\, \mathrm{tr}[FF]} \equiv \left\langle \prod_{f=1}^{N_f} \prod_n (i\lambda_n + m_f) \right\rangle_{N_f=0}, \tag{2.37}$$

suggests that there should be a condensation of small Dirac eigenvalues. Otherwise the effect of small masses would be completely overwhelmed by large Dirac eigenvalues. The time-honored relation due to Banks and Casher [172] indeed reveals a direct connection between the spectral density of Dirac eigenvalues and the chiral condensate *in the chiral limit*:

$$\langle \bar{\psi}\psi \rangle = - \lim_{\lambda\to 0} \lim_{m\to +0} \lim_{V_4\to\infty} \frac{\pi}{V_4} \rho(\lambda; m)\,, \tag{2.38}$$

where

$$\rho(\lambda; m) \equiv \left\langle \sum_n \delta(\lambda - \lambda_n) \right\rangle_{N_f} \tag{2.39}$$

defines the Dirac spectral density with respect to the weight (2.36) and the LHS stands for the chiral condensate in the chiral and infinite-volume limit. The meaning of $\rho$ is such that the quantity $\rho(\lambda; m)\, d\lambda$ counts the number of eigenvalues contained in the interval $[\lambda, \lambda + d\lambda]$. We have $\rho(\lambda) = \rho(-\lambda)$ owing to the chiral symmetry.

In this formula the order of limits is essential. If we take the chiral limit *before* the thermodynamic limit, one finds $\rho(\lambda \to 0) = 0$, meaning that the symmetry is never broken spontaneously in a finite volume. We call $\lim_{V_4\to\infty} \rho(\lambda; m)/V_4$ **the macroscopic spectral density**, to distinguish it from another limit of $\rho$ defined later.

Since this relation is important in many of the later analyses in this thesis, let me review its derivation briefly. First of all, let $\nu \in \mathbb{Z}$ denote the number of exact zero modes of the Dirac operator (i.e., the winding number) in a given gauge configuration. Then

$$- \int d^4x\, \langle \bar{\psi}_f \psi_f \rangle = \frac{\partial}{\partial m_f} \log Z(\{m_g\}) \tag{2.40}$$

$$= \frac{1}{Z(\{m_g\})} \frac{\partial}{\partial m_f} \left\langle \prod_{g=1}^{N_f} \left\{ m_g^{|\nu|} \prod_{\lambda_n>0} (\lambda_n^2 + m_g^2) \right\} \right\rangle_{N_f=0} \tag{2.41}$$

$$= \frac{\langle |\nu| \rangle_{N_f}}{m_f} + \left\langle \sum_{\lambda_n>0} \frac{2m_f}{\lambda_n^2 + m_f^2} \right\rangle_{N_f}, \tag{2.42}$$

$$\therefore\ - \langle \bar{\psi}\psi \rangle = \frac{\langle |\nu| \rangle_{N_f}}{V_4 m} + \frac{1}{V_4} \int_0^\infty d\lambda\, \frac{2m}{\lambda^2 + m^2} \rho(\lambda; m)\,, \tag{2.43}$$

where we dropped the subscript $f$ in the last line for simplicity. Let us take the thermodynamic limit with $m$ fixed. Owing to the central limit theorem $\langle \nu^2 \rangle \sim V_4$, so that the first term disappears in this limit. In the second term, one can use $\lim_{m\to+0} \frac{2m}{\lambda^2+m^2} = \pi\delta(\lambda)$ to show

$$\lim_{m\to+0}\lim_{V_4\to\infty} \langle \bar{\psi}\psi \rangle = -\pi \int_0^\infty d\lambda\, \delta(\lambda) \lim_{m\to+0}\lim_{V_4\to\infty} \frac{\rho(\lambda; m)}{V_4} \tag{2.44}$$

$$= -\pi \lim_{\lambda\to 0}\lim_{m\to+0}\lim_{V_4\to\infty} \frac{\rho(\lambda; m)}{V_4}, \tag{2.45}$$

which completes the proof. We can give a meaning to this relation even for $V_4 < \infty$ if the condition $V_4|\langle\bar{\psi}\psi\rangle|m \gg 1$ is satisfied. To see this, first note that $\rho(0) \sim V_4|\langle\bar{\psi}\psi\rangle|$ indicates that the eigenvalue spacing in the low-end of the spectrum is roughly given by $\frac{1}{V_4|\langle\bar{\psi}\psi\rangle|}$. If the above inequality is satisfied, the eigenvalues are densely distributed near zero, and one can give a well-defined meaning to the spectral density; namely it makes sense to replace the sum in (2.40) by the integral (2.42). Therefore tendency toward the chiral symmetry breaking can be observed even in a finite volume provided $V_4|\langle\bar{\psi}\psi\rangle|m \gg 1$, as pointed out by Leutwyler and Smilga [151].

The above derivation, especially the step from (2.42) to (2.44), assumes that $m$ is much shorter than the typical scale on which $\rho(\lambda; m)$ varies substantially. It is instructive to see why this condition is necessary. Suppose we have

$$\lim_{V_4\to\infty} \frac{\rho(\lambda; m)}{V_4} = \begin{cases} \Sigma\lambda/m & \text{for } 0<\lambda<m, \\ \Sigma & \text{for } m \le \lambda, \end{cases} \tag{2.46}$$

for some constant $\Sigma > 0$. It then follows that

$$\int_0^\infty d\lambda \frac{2m}{\lambda^2+m^2} \lim_{V_4\to\infty} \frac{\rho(\lambda; m)}{V_4} = \left(\frac{\pi}{2} + \log 2\right)\Sigma, \tag{2.47}$$

for arbitrary $m$. This value is not equal to $\lim_{\lambda\to 0}\lim_{m\to+0}\lim_{V_4\to\infty} \pi\frac{\rho(\lambda; m)}{V_4} = \pi\Sigma$. The point is that the replacement

$$\lim_{m\to+0}\left(\frac{2m}{\lambda^2+m^2}\lim_{V_4\to\infty}\frac{\rho(\lambda; m)}{V_4}\right) \to \left(\lim_{m\to+0}\frac{2m}{\lambda^2+m^2}\right)\left(\lim_{m\to+0}\lim_{V_4\to\infty}\frac{\rho(\lambda; m)}{V_4}\right) \tag{2.48}$$

is not necessarily correct. Because the function $\frac{2m}{\lambda^2+m^2}$ (when integrated over $\lambda$) receives the dominant contribution from the region $\lambda \approx m$, it follows that the function $\lim_{V_4\to\infty}\frac{\rho(\lambda; m)}{V_4}$ must be constant over this region, for (2.48) to be valid. This condition is clearly violated in the above counter-example (2.46). In reality, it is known that the Dirac spectral density does not exhibit such a pathology, and the Banks-Casher relation is confirmed on the lattice with a proper account of finite volume effects [158].

Why such a condensation of zero modes occurs is an intrinsically dynamical problem. One of the scenarios argued in the literature states that the QCD vacuum is a liquid of instantons; the exact zero modes associated with each (anti-)instanton are slightly lifted owing to the their overlap, resulting in a nonzero spectral density near zero (see [173, 174] for reviews). However there seems to be no definitive consensus about the origin of ChSB.

Since the work of Banks and Casher, there has been plenty of works regarding the spectral density. It was Smilga and Stern [175] who was the first to derive the form of the macroscopic spectral density for $\lambda > 0$ in the chiral limit. Their result reads

$$\lim_{m\to+0}\lim_{V_4\to\infty}\frac{\rho(\lambda;m)}{V_4} = -\frac{1}{\pi}\langle\bar{\psi}\psi\rangle + \frac{\langle\bar{\psi}\psi\rangle^2(N_f^2-4)}{32\pi^2 N_f F_\pi^4}|\lambda| + o(\lambda)\,. \tag{2.49}$$

Since the LHS must be symmetric for $\lambda \leftrightarrow -\lambda$ this result shows that the macroscopic spectral density has a cusp-like singularity at $\lambda = 0$ for $N_f > 2$. Precisely speaking, this result is valid in the regime

$$\frac{1}{V_4|\langle\bar{\psi}\psi\rangle|} \ll m \ll \lambda \ll \Lambda_{\text{QCD}} \qquad \text{with } \frac{1}{\Lambda_{\text{QCD}}} \ll L\,. \tag{2.50}$$

The second condition is necessary to suppress the contribution of heavy hadrons (with masses $\sim \Lambda_{\text{QCD}}$) to the partition function: $\mathrm{e}^{-L\Lambda_{\text{QCD}}} \ll 1$.

To derive (2.49), Smilga and Stern first calculated the infrared singularity of the scalar susceptibility

$$K^{ab}(m) \equiv \int d^4x \int d^4y\,\langle\bar{\psi}t^a\psi(x)\bar{\psi}t^b\psi(y)\rangle \qquad \{t^a\} = \mathfrak{su}(N_f) \tag{2.51}$$

in the limit $m \to 0$ by using ChPT, and then deduced the existence of the term $\propto |\lambda|$ in (2.49) based on the spectral representation of $K^{ab}(m)$. (See [176] for an alternative derivation.)

On the other hand, for $\lambda \gg \Lambda_{\text{QCD}}$ the Dirac spectrum is governed by the perturbation theory rather than ChPT, resulting in the asymptotic behavior

$$\lim_{V_4\to\infty}\frac{\rho(\lambda;m)}{V_4} \sim \frac{N_c}{4\pi^2}\lambda^3\,. \tag{2.52}$$

We do not precisely know how these distinct behaviors of the macroscopic spectral density are related to each other at $\lambda \approx \Lambda_{\text{QCD}}$. We note also that this part of the macroscopic spectral density does not affect the derivation of the Banks-Casher relation, as long as the UV cutoff (such as the lattice spacing $a$) is removed *after* the chiral limit. The correct order of limits is, therefore, '**First**, the thermodynamic limit $V_4 \to \infty$. **Secondly**, the chiral limit $m \to 0$. And **Thirdly**, the continuum limit $a \to 0$.'

Later a more systematic method of obtaining the Dirac spectral density was developed [177–179] based on the partially quenched chiral perturbation theory (pqChPT) [180, 181]. (This method is also called the supersymmetric method, although it has nothing to do with the spacetime supersymmetry.) The extension of (2.49) to the case of nonzero sea quark masses was achieved in this framework [177]. (See [156, 182] for reviews.) Below we outline the essential idea of this method, since it becomes necessary in the next section where we identify the relationship between ChPT and ChRMT.

Let us consider the partition function of QCD with $N_f$ flavors of fermionic quarks of mass $m$, one fermionic quark of mass $m_v$ and one *bosonic* quark of mass $m'_v$:

$$Z_{N_f+1|1}(m, m_v|m'_v) = \left\langle \det{}^{N_f}(D+m)\frac{\det(D+m_v)}{\det(D+m'_v)} \right\rangle_{N_f=0}. \tag{2.53}$$

Since the bosonic quark does not obey the theorem of spin and statistics, it cannot be seen as a physical entity; it should be regarded as a mathematical device. Then

$$\left\langle \mathrm{tr}\frac{1}{D+m_v} \right\rangle_{N_{f+1|1}} = \frac{\partial}{\partial m_v}\log Z_{N_{f+1|1}}(m, m_v|m'_v). \tag{2.54}$$

By putting $m_v = m'_v$ on both sides, we can erase the ratio of determinants in (2.53). The result is therefore

$$\left\langle \mathrm{tr}\frac{1}{D+m_v} \right\rangle_{N_f} = \frac{\partial}{\partial m_v}\log Z_{N_f+1|1}(m, m_v|m'_v)\bigg|_{m_v=m'_v} \tag{2.55}$$

$$\equiv V_4\Sigma_v(m; m_v). \tag{2.56}$$

The LHS is called the *resolvent* of the Dirac operator. It is nothing but the chiral condensate for a valence quark of mass $m_v$. What is important here is that $m_v$ is now an independent parameter: we can change it while keeping the sea quark mass $m$ fixed. This is the advantage of introducing a bosonic quark in the intermediate step.

The resolvent carries all information on the spectral density. Namely, we have

$$\frac{1}{V_4}\rho(\lambda; m) = \frac{1}{V_4}\left\langle \sum_n \delta(\lambda-\lambda_n) \right\rangle_{N_f} \tag{2.57}$$

$$= -\frac{1}{2\pi}\frac{1}{V_4}\lim_{\varepsilon\to+0}\left\langle \sum_n \left( \frac{1}{-i(\lambda-\lambda_n)-\varepsilon} - \frac{1}{-i(\lambda-\lambda_n)+\varepsilon} \right) \right\rangle_{N_f} \tag{2.58}$$

$$= -\frac{1}{2\pi}\frac{1}{V_4}\lim_{\varepsilon\to+0}\left\langle \mathrm{tr}\frac{1}{D-(i\lambda+\varepsilon)} - \mathrm{tr}\frac{1}{D-(i\lambda-\varepsilon)}\right\rangle_{N_f} \tag{2.59}$$

$$= -\frac{1}{2\pi}\lim_{\varepsilon\to+0}\left\{\Sigma_v(m;-i\lambda-\varepsilon) - \Sigma_v(m;-i\lambda+\varepsilon)\right\}. \tag{2.60}$$

Similar formulae exist also for general $k$ point spectral correlation functions ($k > 1$), which involves $Z_{N_f+k|k}$ instead of $Z_{N_f+1|1}$. Therefore the problem of computing the spectral density boils down to the question of valence quark mass dependence of chiral condensate. It follows if we can compute the valence quark mass dependence of the partition function $Z_{N_f+k|k}$. For $m_v \ll \Lambda_{\mathrm{QCD}}$ this can be achieved by ChPT, but a novel feature as compared to the usual ChPT (2.20) appears: the pattern of chiral symmetry breaking is

$$\mathrm{Gl}(N_f+k|k)_R \otimes \mathrm{Gl}(N_f+k|k)_L \to \mathrm{Gl}(N_f+k|k)_V, \tag{2.61}$$

which involves graded groups (also known as 'supergroups'). (There is a subtlety in the correct identification of the coset; see the original references.) Approximating the partition function by the theory of pions originating from (2.61), one can compute $\Sigma_v$ and $\rho(\lambda; m)$ order by order in the low-energy expansion.

## *2.3.2 The ε-Regime of ChPT*

It has been known for long that the usual "$p$-expansion" (2.21) gets in trouble when the chiral limit is approached in a fixed volume. According to this scheme we must have $V_4|\langle\bar\psi\psi\rangle|m = \mathcal{O}(1/p^2) \gg 1$,[9] implying that the perturbative expansion loses its convergence when $m$ is so small that $V_4|\langle\bar\psi\psi\rangle|m = \mathcal{O}(1)$, which is called the $\varepsilon$-regime of QCD [151, 183]. The origin of infrared singularity in the $p$-expansion in the chiral limit can be easily seen in the following way. When we expand the chiral Lagrangian (2.20) around the tree-level vacuum $U = \mathbf{1}$ to second order in the pion field $U = \exp(i\pi(x)/F_\pi)$, we find the kinetic term $\sim (\partial_\mu\pi)^2 + (\Sigma_0 m/F_\pi^2)\pi^2$. Since in the chiral limit $m \to 0$ the zero-momentum modes disappear from this expression, the integral for zero modes is not Gaussian, and we have to treat the zero mode as a *collective coordinate*:

$$U(x) \equiv U\exp(i\xi(x)/F_\pi), \qquad U : \text{zero mode}, \quad \xi(x) : \text{nonzero mode}. \tag{2.62}$$

Gasser and Leutwyler [183] proposed to reorganize the perturbative expansion based on what is called the $\varepsilon$-counting (see Table 2.4). In this scheme *all* Feynman diagrams that exclusively contain zero modes are on the same order. They are summed up exactly, after which the expansion suffers from no infrared singularity at all. At the leading order of the $\varepsilon$-expansion, the QCD partition function becomes

---

[9] $\langle\bar\psi\psi\rangle$ is inserted to make the combination dimensionless.

**Table 2.4** Comparison of two expansions in ChPT. The counting of the chemical potential $\mu$ is included for completeness.

- $p$-expansion [167]
  - — Counting: $\partial_\mu \sim \xi(x) \sim 1/L \sim T \sim \mu \sim \mathcal{O}(p)$, $m \sim \mathcal{O}(p^2)$
  - — Rapid convergence if $V_4|\langle\bar{\psi}\psi\rangle|m \gg 1$
- $\varepsilon$-expansion [151, 183]
  - — Counting: $\partial_\mu \sim \xi(x) \sim 1/L \sim T \sim \mathcal{O}(\varepsilon)$, $\mu \sim \mathcal{O}(\varepsilon^2)$, $m \sim \mathcal{O}(\varepsilon^4)$
  - — Rapid convergence if $m_\pi L \ll 1$

(There also exist other schemes, e.g., the $\delta$-regime [184] and the intermediate expansion proposed by Damgaard and Fukaya [185])

$$Z(M) = \int_{\mathrm{SU}(N_f)} [dU] \int [d\xi] \exp\left( V_4 \Sigma_0 \operatorname{Re}\operatorname{tr}[MU] - \int d^4x \, \operatorname{tr}[(\partial_\mu \xi)^2] \right) \times [1 + \mathcal{O}(\varepsilon^2)], \tag{2.63}$$

where $[dU]$ stands for the Haar measure of $\mathrm{SU}(N_f)$.

If we take the limit $V_4 \to \infty$ and $m \to 0$ with fixed $V_4 \Sigma_0 m$, the leading order expression becomes *exact*, and remarkably the quark mass dependence of the partition function is simply given as

$$Z(M) = \int_{\mathrm{SU}(N_f)} [dU] \exp\left(V_4 \Sigma_0 \operatorname{Re}\operatorname{tr}[MU]\right). \tag{2.64}$$

The infinite-dimensional path-integral has now reduced to an ordinary integral which one can calculate explicitly! Without invoking a systematic $\varepsilon$-expansion one can also reach this expression in the following way: when the condition

$$\frac{1}{\Lambda_{\mathrm{QCD}}} \ll L \ll \frac{1}{m_\pi} \tag{2.65}$$

is satisfied, contributions of heavier hadrons to the partition function is negligible, while the pion zero modes dominate the other modes with nonzero momentum, as is clear from the propagator

$$G_\pi(x) = \frac{1}{V_4} \sum_{n_\mu \in \mathbb{Z}^4} \frac{e^{i x_\mu n_\mu / L}}{m_\pi^2 + (n_\mu/L)^2} \simeq \frac{1}{V_4 m_\pi^2}. \qquad (\because m_\pi \ll 1/L\,.) \tag{2.66}$$

Therefore the kinetic term in the chiral Lagrangian can be simply dropped, yielding (2.64).

This scheme can be extended to the partially quenched case (i.e., in the presence of bosonic quarks), enabling us to determine the valence quark mass dependence of the chiral condensate. Based on the resolvent method reviewed in the previous section, one can then derive the spectral density in the $\varepsilon$-regime. This is called **the**

**microscopic spectral density**, which we denote by $\rho_s$ to distinguish it from the macroscopic spectral density $\rho$. To summarize,

- The *macroscopic* spectral density $\rho(\lambda)$ can be derived from pqChPT in the $p$-regime, for $\frac{1}{V_4\Sigma_0} \ll m, \lambda \ll \Lambda_{\mathrm{QCD}}$.
- The *microscopic* spectral density $\rho_s(\lambda)$ can be derived from pqChPT in the $\varepsilon$-regime, for $m, \lambda \ll \frac{F_\pi^2}{\sqrt{V_4\Sigma_0}} \equiv E_T$.[10]

The threshold energy $E_T$ is called the Thouless energy in relation to mesoscopic physics. In the domain $\dfrac{1}{V_4\Sigma_0} \ll \lambda \ll \dfrac{F_\pi^2}{\sqrt{V_4\Sigma_0}}$ where both expansions are applicable, it was confirmed that they indeed give identical results [177]. See Fig. 2.3 for a schematic illustration of the Dirac spectrum.

It was Leutwyler and Smilga who showed that the $\varepsilon$-regime partition function (2.64) (without bosonic quarks) is useful to gain insight into the effect of the gauge field topology on the microscopic spectral correlation functions [151].[11] Let us outline their work. As is well known, gauge field configurations with finite action can be classified into sectors with different winding numbers $\nu \in \mathbb{Z}$. The QCD vacuum is a superposition of those states, the so-called $\theta$-vacua, which is characterized by a parameter $\theta$ ($\approx 0$ in reality). Taking this into account, we have in place of (2.64)

$$Z(\theta, M) = \int_{\mathrm{SU}(N_f)} [dU] \exp\left(V_4\Sigma_0 \,\mathrm{Re}\,\mathrm{tr}[\mathrm{e}^{i\theta/N_f} MU]\right) \tag{2.67}$$

$$\equiv \sum_{\nu=-\infty}^{\infty} \mathrm{e}^{i\nu\theta} Z_\nu(M), \tag{2.68}$$

$$Z_\nu(M) \equiv \int_{\mathrm{U}(N_f)} [dU]\, (\det U)^\nu \exp\left(V_4\Sigma_0 \,\mathrm{Re}\,\mathrm{tr}[MU^\dagger]\right) \tag{2.69}$$

$$\propto (\det M)^\nu \left(1 + \frac{(V_4\Sigma_0)^2}{4(N_f+|\nu|)} \mathrm{tr}[M^\dagger M] + \mathcal{O}(M^4)\right). \quad [\nu \geq 0] \tag{2.70}$$

For $\nu < 0$ the prefactor is to be replaced by $(\det M^\dagger)^{-\nu}$.
On the other hand, the spectral representation of the partition function reads

$$Z_\nu(M) = \int [dA_\mu]_\nu \det\left(D + \frac{1-\gamma_5}{2}M + \frac{1+\gamma_5}{2}M^\dagger\right) \mathrm{e}^{-\frac{1}{2}\int d^4x\, \mathrm{tr}[FF]} \tag{2.71}$$

---

[10] This condition follows from (2.65) applied to the pion mass composed of valence quarks.

[11] Here we are reviewing the stuff backward in time, since [151] was done before [177, 178].

$$\propto (\det M)^{\nu} \left\langle \prod_{\lambda_n > 0} \det(\lambda_n^2 + MM^{\dagger}) \right\rangle_{N_f=0,\,\nu} \tag{2.72}$$

$$\propto (\det M)^{\nu} \left\langle \prod_{\lambda_n > 0} \det\left(1 + \frac{MM^{\dagger}}{\lambda_n^2}\right) \right\rangle_{N_f,\nu} \tag{2.73}$$

where $\langle \dots \rangle_{N_f,\nu}$ denotes the average taken over gauge field configurations with a fixed topology $\nu$ and $N_f$ flavors of *massless* quarks.

By comparing the IR expression (2.70) with the UV expression (2.71) order by order, Leutwyler and Smilga derived novel spectral sum rules for the microscopic Dirac eigenvalues, e.g.,

$$\left\langle \sum_{\ell_n > 0} \frac{1}{\ell_n^2} \right\rangle_{N_f,\nu} = \frac{1}{4(N_f + |\nu|)}. \qquad [\ell_n \equiv V_4 \Sigma_0 \lambda_n] \tag{2.74}$$

This is consistent with the indication from the Banks-Casher relation that the scale of smallest Dirac eigenvalues is set by $\frac{1}{V_4 \Sigma_0}$. This is the characteristic scale in the $\varepsilon$-regime. Comparing this to a naïve perturbative estimate $\sim 1/L$ one recognizes that the strong gauge interaction pushes nonzero eigenvalues toward the origin substantially.

Such spectral sum rules give us far more information than the Banks-Casher relation about spectral correlations of smallest Dirac eigenvalues: e.g., one can learn the effect of nonzero $\nu$, one can infer the behavior of near-zero modes when the thermodynamic limit is approached, etc. *This way one can see a smooth crossover from a small volume where chiral condensate vanishes to a large volume where a nonzero density* $\rho(0)/V_4 \neq 0$ *develops.*

There are a few supplementary remarks. **(I)** These sum rules can also be derived for $N_f = 1$, for which the chiral condensate is not an order parameter of symmetry breaking and no pions appear [151]. **(II)** All spectral sum rules (including (2.74)) can be generalized to nonzero sea quark masses ($V_4 \Sigma_0 m \neq 0$). They are called the *massive* spectral sum rules [186]. **(III)** The sum rules are based on the leading order of the $\varepsilon$-expansion. In realistic cases with $V_4 < \infty$ and $m > 0$, there are corrections in general. **(IV)** The original analysis of Leutwyler and Smilga [151] which was mostly restricted to fermions in the complex representation of the gauge group was generalized by Smilga and Verbaarschot [115] to fermions in real and pseudoreal representations, which exhibit SSB patterns different from the usual one (see Table 2.2). For pseudoreal fermions (pertinent to two-color QCD), the partition function in the $\varepsilon$-regime readily follows from the mass term in (2.25):

$$Z(\theta; m) = \int_{\mathrm{SU}(2N_f)/\mathrm{Sp}(2N_f)} [d\Sigma] \exp\left(V_4 G m \,\mathrm{Re}\,\mathrm{tr}[e^{i\theta/N_f} I_{2N_f} \Sigma]\right), \tag{2.75}$$

The massless spectral sum rules are obtained by expanding the $\nu$-th component $Z_\nu$ of $Z(\theta; m)$ in powers of $m$ and matching with the UV expression thereof. The result reads

$$\left\langle \sum_{\ell_n>0} \frac{1}{\ell_n^2} \right\rangle_{N_f,\nu} = \frac{1}{2N_f - 1 + |\nu|}. \qquad [\ell_n \equiv V_4 G \lambda_n] \tag{2.76}$$

How restrictive are those spectral sum rules? While the microscopic spectral correlation functions are verified to reproduce the Leutwyler-Smilga sum rules, the converse is not true. It was explicitly shown by Verbaarschot using ChRMT [187] that different spectral densities satisfy identical spectral sum rules for $N_f = 1$ [187]. Another counter-example occurs in QCD with nonzero chemical potential, which will be reviewed later. The bottom line is that *we must introduce bosonic quarks in order to derive not only sum rules but also the full spectral correlation functions.*

However, there is a useful method that gives same results as those of pqChPT in the leading order of the $\varepsilon$-expansion. The method, called the chiral random matrix theory, is mathematically far easier to handle than pqChPT. We will review it in the next section.

### 2.3.3 Chiral Random Matrix Theory

Random matrix theory (RMT) has found numerous applications within and outside of physics, e.g., in nuclear physics [188], mesoscopic condensed matter physics [189, 190], quantum gravity [191], number theory [192], econophysics [193], wireless communications [194], and many more; see [195, 196] for books.

Underlying the applicability of RMT to diverse quantum systems is an empirical fact that *statistical fluctuations of energy levels on the scale of one average level spacing are universal, in the sense that they are solely determined by symmetries of the system.* A proposed condition for RMT to be applicable to a specific quantum system is nowadays called as the BGS conjecture, which posits that the energy level correlations of a quantum system should be described by RMT when the classical limit of the system is chaotic [197]. There have been numerous studies on this topic; see e.g., [198] and references therein for recent developments.

In comparing physical energy levels with those from RMT it is mandatory to perform a procedure called *unfolding* on the physical spectra, by which its average level spacing becomes unity everywhere on the spectrum. More explicitly we transform the dimensionful $\{E_n\}_n$ to dimensionless $\{\ell_n\}_n$ using the formula

$$\ell_n \equiv \int_0^{E_n} dE\, \rho(E), \tag{2.77}$$

where $\rho(E)$ is the average spectral density. Let us now consider the Dirac operator spectra $\{\lambda_n\}_n$ near zero-virtuality in QCD. According to the Banks-Casher relation

(2.38) we have $\rho(0) = V_4|\langle\bar{q}q\rangle|$, hence

$$\ell_n = \int_0^{\lambda_n} d\lambda\, \rho(\lambda) \tag{2.78}$$
$$\approx \rho(0)\lambda_n = V_4|\langle\bar{q}q\rangle|\lambda_n. \tag{2.79}$$

Interestingly this dimensionless combination is what appears in the Leutwyler-Smilga spectral sum rules reviewed in the previous section. It is therefore natural to ask if the distribution of unfolded Dirac eigenvalues $\{\ell_n\}_n$ are universal. This question was answered in the affirmative by Verbaarschot and his collaborators [152, 199, 200], who showed that the distribution of near-zero Dirac eigenvalues are indeed universal, in the sense that it is solely determined by chiral symmetry and its spontaneous breaking in QCD. Furthermore it followed that spectral correlations in QCD Dirac spectra may be derived from a RMT with the correct symmetries of QCD but is simple enough to solve analytically. This is called chiral random matrix theory (ChRMT). Since then it has been extended and consolidated significantly (see [201–205] for reviews).

In this section we summarize basic features and main achievements of ChRMT. ChRMT is a zero-dimensional model of QCD with the same global symmetries as QCD. The partition function of ChRMT for QCD at $T = \mu = 0$ is defined by

$$Z_{N_f,\nu}(\{m_f\}) = \int \prod_{i,j} dW_{ij} \prod_{f=1}^{N_f} \det(\mathcal{D} + m_f \mathbf{1}_{2N+\nu})\mathrm{e}^{-N\alpha^2 \operatorname{tr}[WW^\dagger]} \tag{2.80}$$

where $\alpha$ is a numerical constant and

$$\mathcal{D} = \begin{pmatrix} \mathbf{0} & W \\ -W^\dagger & \mathbf{0} \end{pmatrix}, \tag{2.81}$$

which mimics the Dirac operator in QCD with chiral symmetry: $\{\gamma_5, D\} = 0$. $W$ is a matrix of size $N \times (N + \nu)$. Then the $2N + \nu$ eigenvalues of $\mathcal{D}$ appear in $N$ pairs with opposite sign $(\lambda, -\lambda)$ plus $|\nu|$ zero eigenvalues, implying that $\nu$ corresponds to the topological charge of gauge fields in QCD. There exist three ensembles in ChRMT, corresponding to three classes of chiral symmetry in QCD in Table 2.2:

- $\beta = 1 \cdots$ **Chiral Gaussian Orthogonal Ensemble (ChGOE)**
  Elements of $W$: *real*, ChSB: $\mathrm{SU}(2N_f) \to \mathrm{Sp}(2N_f)$
- $\beta = 2 \cdots$ **Chiral Gaussian Unitary Ensemble (ChGUE)**
  Elements of $W$: *complex*, ChSB: $\mathrm{SU}(N_f) \times \mathrm{SU}(N_f) \to \mathrm{SU}(N_f)$
- $\beta = 4 \cdots$ **Chiral Gaussian Symplectic Ensemble (ChGSE)**
  Elements of $W$: *quaternion real*, ChSB: $\mathrm{SU}(2N_f) \to \mathrm{SO}(2N_f)$

The parameter $\beta$, called the Dyson index, counts the number of degrees of freedom per each matrix element. The three classes are sometimes called the *three-fold way* in QCD, borrowing the terminology for the classical Dyson-Wigner ensembles (GOE, GUE and GSE) [206].[12] A similar classification is known for QCD in three-dimensions [209].

Their connection to QCD shows up in the large-$N$ limit of the partition function (2.80). The large-$N$ limit can be taken in 5 steps:

(1) Rewrite the determinants as exponentials by introducing auxiliary Grassmann variables,
(2) Integrate out Gaussian matrix elements, which yields four-fermion terms,
(3) Bilinearize the four-fermion terms by a Hubbard-Stratonovich transformation,
(4) Integrate out Grassmann variables,
(5) Take the large-$N$ limit in the saddle-point approximation.

The result, for the $\beta = 2$ class, reads [199]

$$Z_{N_f,\nu}(\{m_f\}) \sim \int_{\mathrm{U}(N_f)} [dU]\,(\det U)^{\nu} \exp[N\alpha \,\mathrm{Re}\,\mathrm{tr}[MU]]. \tag{2.82}$$

This is identical to the finite-volume partition function in the $\varepsilon$-regime (2.67). Therefore the dimensionless parameters in ChRMT can be related to physical masses and eigenvalues in QCD as follows:

$$N\alpha m_f\Big|_{\mathrm{ChRMT}} \iff V_4\Sigma_0 m_f\Big|_{\mathrm{QCD}}, \tag{2.83}$$

$$N\alpha\lambda_n\Big|_{\mathrm{ChRMT}} \iff \ell_n \equiv V_4\Sigma_0\lambda_n\Big|_{\mathrm{QCD}}, \tag{2.84}$$

where $\{\lambda_n\}_n$ are the eigenvalues of $\mathcal{D}$ in (2.81). The identical dependence of the partition functions on quark masses has an important consequence: *all spectral sum rules by Leutwyler and Smilga for Dirac eigenvalues in the $\varepsilon$-regime of QCD are satisfied by $\{\lambda_n\}_n$ in ChRMT.* It is an impressive piece of evidence that the microscopic spectral correlation functions

$$\rho_s(\ell_1, \ldots, \ell_n) \equiv \lim_{V_4\to\infty} \frac{1}{(V_4\Sigma_0)^n}\rho\left(\frac{\ell_1}{V_4\Sigma_0}, \ldots, \frac{\ell_n}{V_4\Sigma_0}\right) \tag{2.85}$$

in QCD are universal and can be equally calculable in ChRMT. It was rigorously shown in [210, 211] that the partition function of ChRMT with both fermionic and bosonic quarks is equivalent to that of pqChPT in QCD in the $\varepsilon$-regime. Since the latter is a generating functional of all spectral correlation functions, it proves that the

[12] It was shown by Zirnbauer [207] that there are 10 symmetry classes in total for Hermitian random matrix ensembles, which are in one-to-one correspondence with the Riemannian symmetric superspaces (see [208] for a review). The three of them are the Wigner-Dyson ensembles, another three of them are the chiral ensembles described above, and the other four are the so-called the Bogoliubov-de Gennes ensembles.

microscopic correlation functions obtained from ChRMT are identical to those from pqChPT! This is practically important, because the computations in ChRMT can be done with the help of a number of mathematical techniques developed in the field of RMT over half a century, while the integration over a graded coset in pqChPT is highly complicated and intricate.

A caveat to this equivalence between QCD and ChRMT is that ChRMT is only equivalent to the *leading order* of the $\varepsilon$-expansion in QCD; the contributions of pions with nonzero momentum are not taken into account in ChRMT. While the leading order contribution dominates the partition function in the limit $V_4 \to \infty$ with $V_4 \Sigma_0 m_f$ fixed, it is no longer true in a large but finite volume. Those finite-volume effects are not universal, i.e., they do not map to $1/N$ contributions on the ChRMT side. One has to rely on pqChPT to compute subleading effects.

Our discussion is summarized schematically in Fig. 2.4. In QCD or any other theory such as Instanton Liquid Model (ILM) that exhibits ChSB, the low-energy physics is described by an effective theory of pions (= (pq)ChPT). If the quark mass is so small that the Compton wavelength of pions is much larger than the size of the box, the physics is dominated by pions' zero modes (= the $\varepsilon$-regime of (pq)ChPT). The partition function in this limit is also reproduced by ChRMT, if we take the limits $N \to \infty$ and $m_f \to 0$ with $Nm_f$ fixed. This is the theoretical foundation of universality.

Several supplementary remarks are in order.

- While the Dirac operator spectrum is not directly observable, it can be easily measured in lattice QCD simulations. By fitting analytical formula obtained from ChRMT to lattice data, one can extract the value of low energy constants (such as $\Sigma_0$ and $F_\pi$) with a good precision. Namely, ChRMT allows one to obtain quantities in the *thermodynamic limit* from data obtained in simulations in a *finite volume*. In addition the predictions from ChRMT serve to check new algorithms in simulations, since we already know that ChRMT is the exact theory of QCD in a specific limit.
- In (2.80) the weight of matrix elements is chosen to be Gaussian without loss of generality, because it has been proven that the spectrum of random matrices *in the microscopic limit* is insensitive to details of the probability distribution function [212]. This kind of robustness does not exist for the macroscopic spectrum; e.g., the famous semicircle law due to Wigner is known to hold only for a limited class of distribution functions.
- The Dirac eigenvalues in the "bulk" of the spectrum, i.e., away from the origin, also allows a description in terms of RMT [213]. Chiral symmetry is not relevant in the bulk. The origin of universality in this case is not ChSB; it is suspected to be due to quantum chaos.
- While ChRMT has been reviewed above as an exact theory of QCD in a finite volume, there are more phenomenological applications of ChRMT to the QCD phase diagram [214, 215]. It is similar in spirit to Ginzburg-Landau analyses, in the sense that only symmetries of QCD are taken into account.

Summarizing above, we learned that

1. The way the thermodynamic limit is approached in the low end of the Dirac spectrum is universal.
2. ChRMT is a convenient mathematical tool to calculate statistical correlations of microscopic Dirac eigenvalues.
3. Analytical predictions from ChRMT are useful in extracting low energy constants from data obtained in lattice simulations which is inevitably done in a finite volume.

Finally let us just outline how to compute the spectral correlation functions in ChRMT. At least three strategies are known: the method of (skew-)orthogonal polynomials, the replica trick and the supersymmetric method. The supersymmetric method is technically tantamount to pqChPT; one adds bosonic and fermionic valence quarks to the RMT partition function and derive the resolvent of the Dirac operator by differentiating the partition functions w.r.t. the valence quark mass [211]. On the other hand, the replica trick [216, 217] is technically simpler but is known to give exact results in the $\varepsilon$-regime only when it is combined with the Toda lattice equation [218, 219]. While it reveals the meaning of the ChRMT partition function as a $\tau$-function of the Toda hierarchy, it leaves much to be desired, at least in two aspects. First, it is shown to work only for $\beta = 2$. The relevance of the Pfaff hierarchy to $\beta = 1$ and 4 is conjectured [220–223], but remains to be clarified. Second, the method produces the spectral density correctly but it remains unclear how the higher-point functions could also be derived within this approach.

In the literature of (Ch)RMT, the most popular among three seems to be the method of (skew-)orthogonal polynomials; in this approach one typically proceeds as follows [195]:

(I) Change variables from the matrix elements $W_{ij}$ to the eigenvalues of $\mathcal{D}$, $\{\lambda_n\}_{n=1}^N$. Compute the Jacobian associated with this transformation to finally arrive at the eigenvalue representation of the partition function

$$Z = \int \prod_{n=1}^{N} d\lambda_n \, P_N(\lambda_1, \ldots, \lambda_N), \tag{2.86}$$

where $P_N$ is the joint probability distribution function of $N$ eigenvalues.

(II) Express $P_N$ in terms of orthogonal polynomials ($\beta = 2$) or skew-orthogonal polynomials ($\beta = 1,\ 4$). Use their (skew-)orthogonality to integrate out $N - k$ eigenvalues, which leaves behind the $k$-point correlation function

$$R_N^{(k)}(\lambda_1, \ldots, \lambda_k) \equiv \frac{N!}{(N-k)!} \int \prod_{n=k+1}^{N} d\lambda_n \, P_N(\lambda_1, \ldots, \lambda_N). \tag{2.87}$$

It is known that the resultant $k$-point function can be expressed as a determinant for $\beta = 2$ with a kernel $K_N$

$$R_N^{(k)}(\lambda_1, \ldots, \lambda_k) = \det_{1 \le i, j \le k} \left[ K_N(\lambda_i, \lambda_j) \right], \tag{2.88}$$

and a quaternion determinant (or, equivalently, a pfaffian) for $\beta = 1$ and 4. This beautiful property of (Ch)RMT implies that infinitely many correlation functions are governed by just a single kernel!

(III) Take the microscopic limit:

$$\rho_s^{(k)}(\ell_1, \ldots, \ell_k) \equiv \lim_{N \to \infty} \frac{1}{N^k} R_N^{(k)} \left( \frac{\ell_1}{N}, \ldots, \frac{\ell_k}{N} \right). \tag{2.89}$$

Quite recently, Kieburg and Guhr invented a new method that clarifies the origin of the determinantal structure for $\beta = 2$ and a pfaffian structure for $\beta = 1$ and 4 [224, 225]. It was shown that these structures are rooted in the form of the probability weight $P_N$ and have nothing to do with the (skew-)orthogonal polynomials. Their formulae will be used in a later chapter of this thesis.

### 2.3.4 ChRMT at $\mu \neq 0$

Since the invention of RMT, it has long been the ensembles of Hermitian matrices with real eigenvalues that attracted most attention because of the Hermiticity constraint on a physical Hamiltonian. Mathematical theory of non-Hermitian random matrix ensembles with complex eigenvalues was pioneered by Ginibre [226], who introduced three Gaussian ensembles with real, complex and quaternion matrix elements, respectively, but without any further symmetry constraints. However, at that time it lacked a physical application and seemed to be an academic exercise. The situation has changed dramatically over the last two decades; the non-Hermitian RMT now enjoys a number of fruitful applications, ranging from dissipative quantum maps [227, 228], asymmetric neural networks [229, 230], open quantum systems [231] to flux depinning in superconductors [232, 233] and QCD at finite chemical potential [204]. A classification of non-Hermitian random matrix ensembles was given in [234, 235].

In this section we will briefly review the extension of ChRMT to $\mu \neq 0$. The quark chemical potential $\mu$ was first introduced to ChRMT by Stephanov [236], who defined the model

$$Z_{N_f,\nu}(\mu; \{m_f\}) = \int_{\mathbb{C}} \prod_{i,j} dW_{ij} \prod_{f=1}^{N_f} \det[D(\mu) + m_f \mathbf{1}_{2N}] \mathrm{e}^{-N\,tr[WW^\dagger]}, \tag{2.90}$$

where

$$D(\mu) = \begin{pmatrix} \mathbf{0} & W + \mu \mathbf{1}_N \\ -W^\dagger + \mu \mathbf{1}_N & \mathbf{0} \end{pmatrix} \tag{2.91}$$

mimics the QCD Dirac operator at nonzero quark chemical potential. The effect of $\mu \neq 0$ is to make $D(\mu)$ not anti-Hermitian. The eigenvalues spread over the complex plane, instead of being localized on the imaginary axis. Using this model he gave a solution to the puzzle that quenched lattice simulations predict a 'wrong' behavior for the chiral condensate at nonzero chemical potential. His explanation was that the quenched limit does not correspond to the $n \to 0$ limit of $\det^n(D(\mu)+m)$ but to the $n \to 0$ limit of $|\det(D(\mu)+m)|^{2n}$, i.e., QCD with $n$ quarks and $n$ conjugate quarks. The latter is often called the *phase-quenched* QCD. The discrepancy of lattice results and physical expectation is due to the condensation of a bound state of a quark and a conjugate quark. As a result of this, the chiral condensate at fixed $\mu$ vanishes in the chiral limit, in quenched QCD, similarly to two-color QCD at low $\mu$ where the chiral condensate rotates to a diquark condensate (reviewed in Sect. 2.2.3). This explains why quenched approximation is not a good approximation to QCD at $\mu \neq 0$.

The computation of spectral functions in the microscopic limit for this model, where $N\mu^2$, $Nm_f$ and $N\lambda_n$ are fixed in the large-$N$ limit, is hampered by the $\mu$-term which breaks invariance under a unitary transformation $W \to UWV^\dagger$. Only the quenched spectral density was derived in [219] by a supersymmetric method. Osborn proposed to replace the unit matrix multiplying $\mu$ by an independent random matrix [237]:

$$Z_{N_f,\nu}(\mu;\{m_f\}) = \int_{\mathbb{C}} \prod_{i,j} dP_{ij}\, dQ_{ij} \prod_{f=1}^{N_f} \det[D(\mu)+m_f \mathbf{1}_{2N+\nu}] \mathrm{e}^{-N\alpha^2 \operatorname{tr}[PP^\dagger+QQ^\dagger]}, \tag{2.92}$$

with

$$D(\mu) = \begin{pmatrix} \mathbf{0} & P+\mu Q \\ -P^\dagger+\mu Q^\dagger & \mathbf{0} \end{pmatrix}. \tag{2.93}$$

Since the deformation does not change the global symmetry of the model, it is expected based on universality that both models will give identical microscopic correlations of Dirac eigenvalues. Equations (2.90) and (2.92) are called as the one-matrix model and the two-matrix model, respectively. Osborn solved the model for $\beta = 2$ using the method of biorthogonal polynomials. Later it was shown that same results can be obtained from the replica limit of the Toda lattice equation [238]. Osborn et al. revealed that the sign problem for the unquenched case ($N_f > 0$) is reflected in a domain of strong oscillation in the microscopic spectral density, and argued that the oscillation is essential to obtain a nonzero chiral condensate in the chiral limit [239–241].

The large-$N$ limit of (2.92) with most general chemical potentials $\{\mu_f\}_f$ reads [242]

$$Z_{N_f,\nu}(\{\mu_f\};\{m_f\}) \sim \int_{\mathrm{U}(N_f)} [dU]\, (\det U)^\nu \exp\left(2N\alpha \operatorname{Re}\operatorname{tr}[MU] - N \operatorname{tr}[U,B][U^\dagger,B]\right), \tag{2.94}$$

with $M = \text{diag}(m_f)$ and $B = \text{diag}(\mu_f)$. This form is identical to the zero momentum chiral Lagrangian of QCD [243] if we identify

$$N\alpha m_f\Big|_{\text{ChRMT}} \iff V_4 \Sigma_0 m_f\Big|_{\text{QCD}}, \quad N\mu^2\Big|_{\text{ChRMT}} \iff V_4 F_\pi^2 \mu^2\Big|_{\text{QCD}}. \tag{2.95}$$

Therefore by fitting the lattice data to the exact spectral density of ChRMT one can extract both the chiral condensate and the pion decay constant in the infinite volume. It was successfully done in a quenched simulation [244]. In contrast, checking the unquenched results in ChRMT by lattice QCD has been hampered by a severe sign problem. Exact predictions of ChRMT might be useful to check new algorithms in lattice QCD at nonzero chemical potential.

For a flavor-symmetric chemical potential ($B \propto \mathbf{1}_{N_f}$) the partition function $Z_\nu$ and the spectral sum rules for Dirac eigenvalues are entirely independent of $\mu$. It is a result of the fact that pions do not carry a baryon number. It does not contradict the nontrivial $\mu$-dependence of the microscopic spectral density, since the latter does not derive from the chiral Lagrangian but from the *partially quenched* chiral Lagrangian which includes additional quarks of both fermionic and bosonic statistics.

The two-matrix model (2.92) for $\beta = 4$ was introduced in [245] and was solved in [245, 246]. The point is that the anti-unitary symmetry of the Dirac operator in ChGSE persists in the presence of $\mu \neq 0$. As a result, the theory has no sign problem even at $\mu \neq 0$. The exact solution was compared with the result of unquenched simulations and showed excellent agreement [247]. See [204] for a review. The extension to $\beta = 1$ will be considered in a later chapter.

We finish this section by introducing a mathematical terminology. First of all, any $N \times N$ matrix $J$ can be written as $J = H_1 + iH_2$ with two Hermitian matrices $H_1$ and $H_2$. Suppose the matrices $H_1$ and $H_2$ are identically and independently distributed; e.g., according to $\exp\{-N\,\text{tr}(H_1^2 + H_2^2)\}$. Regarding the eigenvalue distribution of the matrix $H_1 + iuH_2$ with $u$ a real deformation parameter, several distinct regimes are known in the large-$N$ limit [248, 249]:

- The Hermitian limit: $N \to \infty$ with $u = 0$.
  The eigenvalues are localized on the real axis and spaced with an average spacing $\sim \mathcal{O}(1/N)$.
- The limit of *weak non-Hermiticity*: $N \to \infty$ with $u \sim \mathcal{O}(1/\sqrt{N})$.
  The eigenvalues are distributed in the vicinity of the real axis and has a tiny imaginary part of magnitude $\mathcal{O}(1/N)$. Their average spacing along the real axis is still $\mathcal{O}(1/N)$.
- The limit of *strong non-Hermiticity*: $N \to \infty$ with $u \sim \mathcal{O}(1)$.
  The eigenvalues spread over the complex plane. Their average distance is $\mathcal{O}(1/\sqrt{N})$. The special case $u = 1$ is called the limit of *maximal non-Hermiticity*, since the matrix $H_1 + iH_2$ is composed of an equal fraction of Hermitian and anti-Hermitian components.

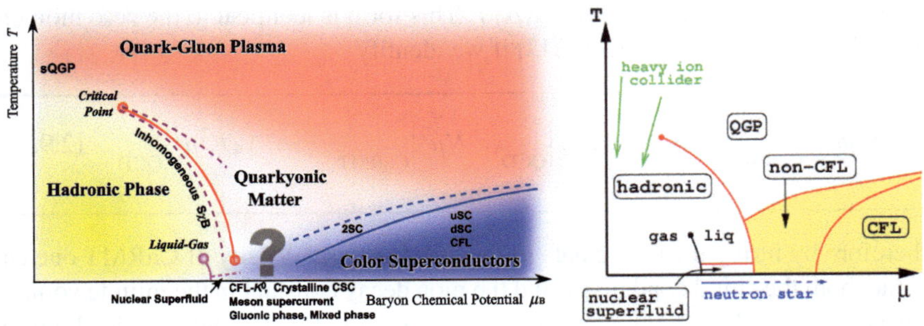


**Fig. 2.1** Conjectured phase diagrams of QCD at finite temperature and baryon chemical potential. (*Left* reprinted from [11] with permission from IOP. *Right* reprinted from [28]. Copyright(2012) by the American Physical Society)

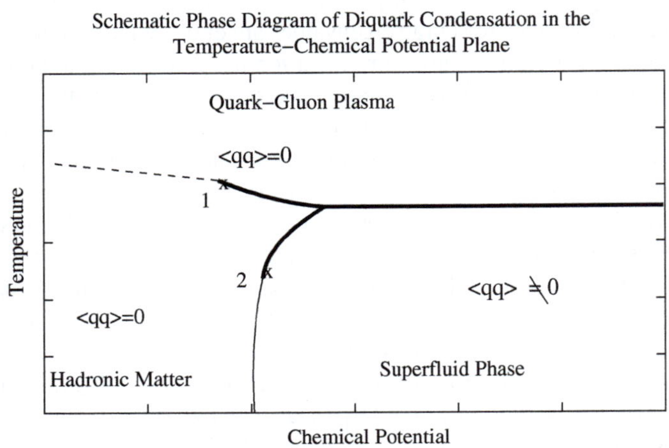


**Fig. 2.2** A schematic phase diagram of two-color QCD for even $N_f$ for $j = 0$ and $m$ small. (Reprinted from [96]. Copyright(2012) with permission from Elsevier). The thick [*thin*] *solid line* denotes the first [second] order phase *transition line*. The *dashed line* denotes a crossover associated with chiral symmetry restoration

It is the limit of weak non-Hermiticity where one can observe a crossover behavior of the eigenvalue distribution between Hermitian and non-Hermitian limits. See [250] for a review.

Recalling that the chemical potential in QCD is scaled as $\mu \sim \mathcal{O}(1/\sqrt{N})$ in (2.95), it corresponds to the limit of weak non-Hermiticity. Then, what is the physical meaning of the limit of strong non-Hermiticity in ChRMT? We will address this question in Chap. 3 Sect. 3.3.

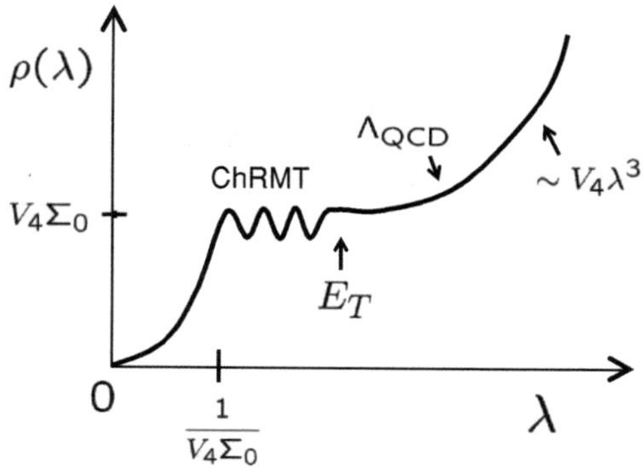


**Fig. 2.3** A schematic landscape of the Dirac spectrum. The smallest eigenvalues $\sim \mathcal{O}(1/V_4\Sigma_0)$ obey the statistics of ChRMT. The interval between $E_T$ and $\Lambda_{\text{QCD}}$ is subject to ChPT in the $p$-regime. The eigenvalues larger than $\Lambda_{\text{QCD}}$ obey perturbative QCD

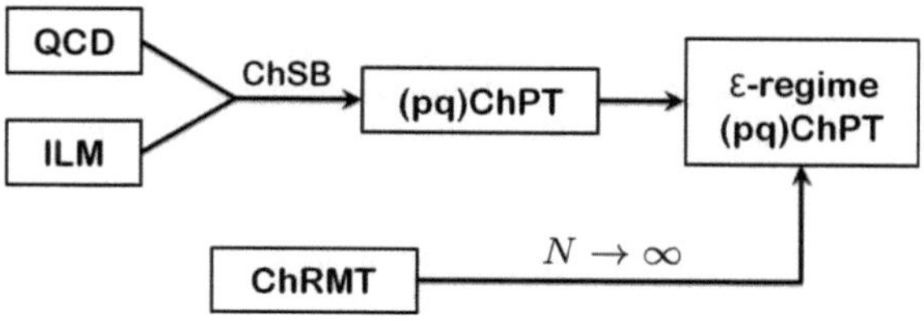


**Fig. 2.4** A schematic illustration of relations between QCD and ChRMT

# References


1. H. Meyer-Ortmanns, Phase transitions in quantum chromodynamics. Rev. Mod. Phys. **68**, 473–598 (1996)
2. A.V. Smilga, Physics of thermal QCD. Phys. Rept. **291**, 1–106 (1997)
3. K. Rajagopal, F. Wilczek, The condensed matter physics of QCD, hep-ph/0011333
4. M.G. Alford, Color superconducting quark matter. Ann. Rev. Nucl. Part. Sci. **51**, 131–160 (2001)
5. M.A. Stephanov, QCD phase diagram: An overview, PoS LAT2006 024 (2006)
6. Q. Wang, Some aspects of color superconductivity: an introduction. Prog. Phys. **30**, 173 (2010)
7. T. Hatsuda, K. Maeda, Quantum Phase Transitions in Dense QCD, ed. by L.D. Carr Chapter 25 in Developments in Quantum Phase Transitions (Taylor and Francis, 2010)
8. P. Braun-Munzinger, J. Wambach, Colloquium: Phase diagram of strongly interacting matter. Rev. Mod. Phys. **81**, 1031–1050 (2009)
9. M. Huang, QCD phase diagram at high temperature and density, arXiv:1001.3216
10. A. Schmitt, Dense matter in compact stars - A pedagogical introduction. Lect. Notes Phys. **811**, 1–111 (2010)
11. K. Fukushima, T. Hatsuda, The phase diagram of dense QCD. Rept. Prog. Phys. **74**, 014001 (2011)
12. J.-Y. Pang, J.-C. Wang, Q. Wang, Some recent progress on quark pairings in dense quark and nuclear matter. Commun. Theor. Phys. **57**, 251 (2012)
13. N.K. Glendenning, *Compact Stars: Nuclear Physics, Particle Physics, and General Relativity* (Springer, New York, 1997), p. 390
14. J.B. Kogut, M.A. Stephanov, The phases of quantum chromodynamics: From confinement to extreme environments. Camb. Monogr. Part. Phys. Nucl. Phys. Cosmol. **21**, 1–364 (2004)
15. K. Yagi, T. Hatsuda, Y. Miake, Quark-gluon plasma: From big bang to little bang. Camb. Monogr. Part. Phys. Nucl. Phys. Cosmol. **23**, 1–446 (2005)

16. P. Haensel, A.Y. Potekhin, D.G. Yakovlev, *Neutron Stars 1: Equation of State and Structure* (Springer, New York, 2007), p. 619
17. K. Splittorff, The sign problem in the epsilon-regime of QCD, PoS LAT2006 023 (2006)
18. S. Ejiri, Recent progress in lattice QCD at finite density, PoS LATTICE2008 002 (2008)
19. P. de Forcrand, Simulating QCD at finite density, PoS LAT2009 010 (2009)
20. O. Philipsen, Lattice QCD at non-zero temperature and baryon density, arXiv:1009.4089.
21. S.P. Klevansky, The Nambu-Jona-Lasinio model of quantum chromodynamics. Rev. Mod. Phys. **64**, 649–708 (1992)
22. T. Hatsuda, T. Kunihiro, QCD phenomenology based on a chiral effective Lagrangian. Phys. Rept. **247**, 221–367 (1994)
23. M. Buballa, NJL model analysis of quark matter at large density. Phys. Rept. **407**, 205–376 (2005)
24. K. Fukushima, Chiral effective model with the Polyakov loop. Phys. Lett. **B591**, 277–284 (2004)
25. K. Fukushima, Phase diagrams in the three-flavor Nambu-Jona-Lasinio model with the Polyakov loop. Phys. Rev. **D77**, 114028 (2008)
26. K. Maeda, G. Baym, T. Hatsuda, Simulating dense QCD matter with ultracold atomic boson-fermion mixtures. Phys. Rev. Lett. **103**, 085301 (2009)
27. J.C. Collins, M.J. Perry, Superdense matter: neutrons or asymptotically free quarks? Phys. Rev. Lett. **34**, 1353 (1975)
28. M.G. Alford, A. Schmitt, K. Rajagopal, T. Schäfer, Color superconductivity in dense quark matter. Rev. Mod. Phys. **80**, 1455–1515 (2008)
29. L.N. Cooper, Bound electron pairs in a degenerate Fermi gas. Phys. Rev. **104**, 1189–1190 (1956)
30. J. Bardeen, L.N. Cooper, J.R. Schrieffer, Microscopic theory of superconductivity. Phys. Rev. **106**, 162 (1957)
31. J. Bardeen, L.N. Cooper, J.R. Schrieffer, Theory of superconductivity. Phys. Rev. **108**, 1175–1204 (1957)
32. M. Tinkham, *Introduction to Superconductivity*, 2nd edn. (Dover Publications, New York, 2004)
33. Y. Nambu, G. Jona-Lasinio, Dynamical model of elementary particles based on an analogy with superconductivity. I. Phys. Rev. **122**, 345–358 (1961)
34. Y. Nambu, G. Jona-Lasinio, Dynamical model of elementary particles based on an analogy with superconductivity. II. Phys. Rev. **124**, 246–254 (1961)
35. S.L. Glashow, Partial symmetries of weak interactions. Nucl. Phys. **22**, 579–588 (1961)
36. S. Weinberg, A model of leptons. Phys. Rev. Lett. **19**, 1264–1266 (1967)
37. A. Salam, *Weak and Electromagnetic Interactions, Originally Printed in Svartholm: Elementary Particle Theory, Proceedings of the Nobel Symposium Held* (At Lerum, Sweden, Stockholm, 1968), pp. 367–377
38. B.C. Barrois, Superconducting quark matter. Nucl. Phys. **B129**, 390 (1977)
39. D. Bailin, A. Love, Superfluidity and superconductivity in relativistic fermion systems. Phys. Rept. **107**, 325 (1984)
40. M.G. Alford, K. Rajagopal, F. Wilczek, QCD at finite baryon density: Nucleon droplets and color superconductivity. Phys. Lett. **B422**, 247–256 (1998)
41. R. Rapp, T. Schafer, E.V. Shuryak, M. Velkovsky, Diquark Bose condensates in high density matter and instantons. Phys. Rev. Lett. **81**, 53–56 (1998)
42. R. Rapp, T. Schafer, E.V. Shuryak, M. Velkovsky, High-density QCD and instantons. Annals Phys. **280**, 35–99 (2000)
43. M. Iwasaki, T. Iwado, Superconductivity in the quark matter. Phys. Lett. **B350**, 163–168 (1995)
44. R. Rapp, T. Schäfer, E.V. Shuryak, M. Velkovsky, Diquark Bose condensates in high density matter and instantons. Phys. Rev. Lett. **81**, 53–56 (1998)
45. D.T. Son, Superconductivity by long-range color magnetic interaction in high-density quark matter. Phys. Rev. **D59**, 094019 (1999)

46. R.D. Pisarski, D.H. Rischke, A first order transition to, and then parity violation in, a color superconductor. Phys. Rev. Lett. **83**, 37–40 (1999)
47. D.K. Hong, An effective field theory of QCD at high density. Phys. Lett. **B473**, 118–125 (2000)
48. D.K. Hong, V.A. Miransky, I.A. Shovkovy, L.C.R. Wijewardhana, Schwinger-Dyson approach to color superconductivity in dense QCD. Phys. Rev. **D61**, 056001 (2000)
49. T. Schafer, F. Wilczek, Superconductivity from perturbative one-gluon exchange in high density quark matter. Phys. Rev. **D60**, 114033 (1999)
50. S. Elitzur, Impossibility of spontaneously breaking local symmetries. Phys. Rev. **D12**, 3978–3982 (1975)
51. F. Sannino, A note on anomaly matching for finite density QCD. Phys. Lett. **B480**, 280–286 (2000)
52. M. Huang, Color superconductivity at moderate baryon density. Int. J. Mod. Phys. **E14**, 675 (2005)
53. M.G. Alford, K. Rajagopal, F. Wilczek, Color-flavor locking and chiral symmetry breaking in high density QCD. Nucl. Phys. **B537**, 443–458 (1999)
54. R. Casalbuoni, R. Gatto, G. Nardulli, Dispersion laws for in-medium fermions and gluons in the CFL phase of QCD. Phys. Lett. **B498**, 179–188 (2001)
55. V.P. Gusynin, I.A. Shovkovy, Collective modes of color-flavor locked phase of dense QCD at finite temperature. Nucl. Phys. **A700**, 577–617 (2002)
56. H. Malekzadeh, D.H. Rischke, Gluon self-energy in the color-flavor-locked phase. Phys. Rev. **D73**, 114006 (2006)
57. T. Hatsuda, M. Tachibana, N. Yamamoto, Spectral continuity in dense QCD. Phys. Rev. **D78**, 011501 (2008)
58. T. Schäfer, Patterns of symmetry breaking in QCD at high baryon density. Nucl. Phys. **B575**, 269–284 (2000)
59. T. Schäfer, F. Wilczek, Continuity of quark and hadron matter. Phys. Rev. Lett. **82**, 3956–3959 (1999)
60. E.H. Fradkin, S.H. Shenker, Phase diagrams of lattice gauge theories with higgs fields. Phys. Rev. **D19**, 3682–3697 (1979)
61. T. Banks, E. Rabinovici, Finite temperature behavior of the lattice abelian higgs model. Nucl. Phys. **B160**, 349 (1979)
62. T. Hatsuda, M. Tachibana, N. Yamamoto, G. Baym, New critical point induced by the axial anomaly in dense QCD. Phys. Rev. Lett. **97**, 122001 (2006)
63. H. Abuki, G. Baym, T. Hatsuda, N. Yamamoto, The NJL model of dense three-flavor matter with axial anomaly: the low temperature critical point and BEC-BCS diquark crossover. Phys. Rev. **D81**, 125010 (2010)
64. H. Basler, M. Buballa, Role of two-flavor color superconductor pairing in a three-flavor Nambu-Jona-Lasinio model with axial anomaly. Phys. Rev. **D82**, 094004 (2010)
65. T. Schafer, Quark hadron continuity in QCD with one flavor. Phys. Rev. **D62**, 094007 (2000)
66. T. Brauner, J. Hosek, R. Sykora, Color superconductor with a color-sextet condensate. Phys. Rev. **D68**, 094004 (2003)
67. M.G. Alford, J. Berges, K. Rajagopal, Unlocking color and flavor in superconducting strange quark matter. Nucl. Phys. **B558**, 219–242 (1999)
68. T. Matsuura, K. Iida, T. Hatsuda, G. Baym, Thermal fluctuations of gauge fields and first order phase transitions in color superconductivity. Phys. Rev. **D69**, 074012 (2004)
69. K. Iida, T. Matsuura, M. Tachibana, T. Hatsuda, Melting pattern of diquark condensates in quark matter. Phys. Rev. Lett. **93**, 132001 (2004)
70. S.D.H. Hsu, F. Sannino, M. Schwetz, Anomaly matching in gauge theories at finite matter density. Mod. Phys. Lett. **A16**, 1871–1880 (2001)
71. T. Schafer, Instantons and the chiral phase transition at non-zero baryon density. Phys. Rev. **D57**, 3950–3961 (1998)
72. E.V. Shuryak, The role of instantons in quantum chromodynamics. 3. Quark - Gluon Plasma. Nucl. Phys. **B203**, 140 (1982)

73. D.T. Son, M.A. Stephanov, A.R. Zhitnitsky, Instanton interactions in dense-matter QCD. Phys. Lett. **B510**, 167–172 (2001)
74. T. Schafer, Instanton effects in QCD at high baryon density. Phys. Rev. **D65**, 094033 (2002)
75. T. Schäfer, QCD and the eta' mass: Instantons or confinement? Phys. Rev. **D67**, 074502 (2003)
76. N. Yamamoto, Instanton-induced crossover in dense QCD. JHEP **12**, 060 (2008)
77. S. Chandrasekharan, A.C. Mehta, Effects of the anomaly on the two-flavor QCD chiral phase transition. Phys. Rev. Lett. **99**, 142004 (2007)
78. J.-W. Chen, K. Fukushima, H. Kohyama, K. Ohnishi, U. Raha, $U_A(1)$ Anomaly in hot and dense QCD and the critical surface. Phys. Rev. **D80**, 054012 (2009)
79. M.G. Alford, A. Kapustin, F. Wilczek, Imaginary chemical potential and finite fermion density on the lattice. Phys. Rev. **D59**, 054502 (1999)
80. D.T. Son, M.A. Stephanov, QCD at finite isospin density: From pion to quark antiquark condensation. Phys. Atom. Nucl. **64**, 834–842 (2001)
81. D.T. Son, M.A. Stephanov, QCD at finite isospin density. Phys. Rev. Lett. **86**, 592–595 (2001)
82. J.B. Kogut, M.A. Stephanov, D. Toublan, J.J.M. Verbaarschot, A. Zhitnitsky, QCD-like theories at finite baryon density. Nucl. Phys. **B582**, 477–513 (2000)
83. R. Slansky, Group theory for unified model building. Phys. Rept. **79**, 1–128 (1981)
84. H. Georgi, Lie algebras in particle physics from isospin to unified theories. Front. Phys. **54**, 1–255 (1982)
85. S. Hands, D.N. Walters, Evidence for BCS diquark condensation in the 3+1d lattice NJL model. Phys. Lett. **B548**, 196–203 (2002)
86. S. Hands, D.N. Walters, Numerical portrait of a relativistic BCS gapped superfluid. Phys. Rev. **D69**, 076011 (2004)
87. A. Nakamura, Quarks and gluons at finite temperature and density. Phys. Lett. **B149**, 391 (1984)
88. S. Hands, J.B. Kogut, M.-P. Lombardo, S.E. Morrison, Symmetries and spectrum of SU(2) lattice gauge theory at finite chemical potential. Nucl. Phys. **B558**, 327–346 (1999)
89. M.-P. Lombardo, Interquark potential, susceptibilities and particle density of two color QCD at finite chemical potential and temperature, hep-lat/9907025
90. S. Hands et al., Numerical study of dense adjoint matter in two color QCD. Eur. Phys. J. **C17**, 285–302 (2000)
91. E. Bittner, M.-P. Lombardo, H. Markum, R. Pullirsch, Dirac and Gorkov spectra in two color QCD with chemical potential. Nucl. Phys. Proc. Suppl. **94**, 445–448 (2001)
92. S. Muroya, A. Nakamura, C. Nonaka, Monte Carlo study of two-color QCD with finite chemical potential: Status report of Wilson fermion simulation. Nucl. Phys. Proc. Suppl. **94**, 469–472 (2001)
93. J.B. Kogut, D. Toublan, D.K. Sinclair, Diquark condensation at nonzero chemical potential and temperature. Phys. Lett. **B514**, 77–87 (2001)
94. J.B. Kogut, D.K. Sinclair, S.J. Hands, S.E. Morrison, Two-colour QCD at non-zero quark-number density. Phys. Rev. **D64**, 094505 (2001)
95. S. Hands, I. Montvay, L. Scorzato, J. Skullerud, Diquark condensation in dense adjoint matter. Eur. Phys. J. **C22**, 451–461 (2001)
96. J.B. Kogut, D. Toublan, D.K. Sinclair, The phase diagram of four flavor SU(2) lattice gauge theory at nonzero chemical potential and temperature. Nucl. Phys. **B642**, 181–209 (2002)
97. B. Alles, M. D'Elia, M.-P. Lombardo, M. Pepe, Vector spectrum and color screening in two color QCD at non-zero T and mu, hep-lat/0210039
98. S. Muroya, A. Nakamura, C. Nonaka, Behavior of hadrons at finite density: Lattice study of color SU(2) QCD. Phys. Lett. **B551**, 305–310 (2003)
99. J.B. Kogut, D. Toublan, D.K. Sinclair, The pseudo-Goldstone spectrum of 2-colour QCD at finite density. Phys. Rev. **D68**, 054507 (2003)
100. S. Muroya, A. Nakamura, C. Nonaka, T. Takaishi, Lattice QCD at finite density: An introductory review. Prog. Theor. Phys. **110**, 615–668 (2003)
101. D.K. Sinclair, J.B. Kogut, D. Toublan, Finite density lattice gauge theories with positive fermion determinants. Prog. Theor. Phys. Suppl. **153**, 40–50 (2004)

102. B. Alles, M. D'Elia, M.P. Lombardo, Behaviour of the topological susceptibility in two colour QCD across the finite density transition. Nucl. Phys. **B752**, 124–139 (2006)
103. S. Hands, S. Kim, J.-I. Skullerud, Deconfinement in dense 2-color QCD. Eur. Phys. J. **C48**, 193 (2006)
104. S. Hands, P. Sitch, J.-I. Skullerud, Hadron spectrum in a two-colour baryon-rich medium. Phys. Lett. **B662**, 405 (2008)
105. M.-P. Lombardo, M.L. Paciello, S. Petrarca, B. Taglienti, Glueballs and the superfluid phase of Two-Color QCD. Eur. Phys. J. **C58**, 69–81 (2008)
106. S. Hands, S. Kim, J.-I. Skullerud, A quarkyonic phase in dense two color matter? Phys. Rev. **D81**, 091502 (2010)
107. S. Hands, T.J. Hollowood, J.C. Myers, Numerical study of the two color attoworld. JHEP **12**, 057 (2010)
108. L.A. Kondratyuk, M.M. Giannini, M.I. Krivoruchenko, The SU(2) color superconductivity. Phys. Lett. **B269**, 139–143 (1991)
109. C. Ratti, W. Weise, Thermodynamics of two-colour QCD and the Nambu Jona-Lasinio model. Phys. Rev. **D70**, 054013 (2004)
110. K. Fukushima, K. Iida, Larkin-Ovchinnikov-Fulde-Ferrell state in two-color quark matter. Phys. Rev. **D76**, 054004 (2007)
111. J.O. Andersen, T. Brauner, Phase diagram of two-color quark matter at nonzero baryon and isospin density. Phys. Rev. **D81**, 096004 (2010)
112. T. Zhang, T. Brauner, D.H. Rischke, QCD-like theories at nonzero temperature and density. JHEP **06**, 064 (2010)
113. L. He, Nambu-Jona-Lasinio model description of weakly interacting Bose condensate and BEC-BCS crossover in dense QCD-like theories. Phys. Rev. **D82**, 096003 (2010)
114. T. Brauner, K. Fukushima, Y. Hidaka, Two-color quark matter: $U(1)_A$ restoration, superfluidity, and quarkyonic phase. Phys. Rev. **D80**, 074035 (2009)
115. A.V. Smilga, J.J.M. Verbaarschot, Spectral sum rules and finite volume partition function in gauge theories with real and pseudoreal fermions. Phys. Rev. **D51**, 829–837 (1995)
116. J.B. Kogut, M.A. Stephanov, D. Toublan, On two-color QCD with baryon chemical potential. Phys. Lett. **B464**, 183–191 (1999)
117. K. Splittorff, D.T. Son, M.A. Stephanov, QCD-like theories at finite baryon and isospin density. Phys. Rev. **D64**, 016003 (2001)
118. J.T. Lenaghan, F. Sannino, K. Splittorff, The superfluid and conformal phase transitions of two-color QCD. Phys. Rev. **D65**, 054002 (2002)
119. K. Splittorff, D. Toublan, J.J.M. Verbaarschot, Diquark condensate in QCD with two colors at next-to-leading order. Nucl. Phys. **B620**, 290–314 (2002)
120. K. Splittorff, D. Toublan, J.J.M. Verbaarschot, Thermodynamics of chiral symmetry at low densities. Nucl. Phys. **B639**, 524–548 (2002)
121. M.A. Metlitski, A.R. Zhitnitsky, theta-parameter in 2 color QCD at finite baryon and isospin density. Nucl. Phys. **B731**, 309–334 (2005)
122. T. Brauner, On the chiral perturbation theory for two-flavor two-color QCD at finite chemical potential. Mod. Phys. Lett. **A21**, 559–570 (2006)
123. T. Kanazawa, T. Wettig, N. Yamamoto, Chiral Lagrangian and spectral sum rules for dense two-color QCD. JHEP **08**, 003 (2009)
124. M. Harada, C. Nonaka, T. Yamaoka, Masses of vector bosons in two-color dense QCD based on the hidden local symmetry. Phys. Rev. **D81**, 096003 (2010)
125. M.B. Hecht, C.D. Roberts, S.M. Schmidt, Diquarks and density. Lect. Notes Phys. **578**, 218–234 (2001)
126. J.J.M. Verbaarschot, Spectrum of the Dirac operator in a QCD instanton liquid: Two versus three colors. Nucl. Phys. **B427**, 534–544 (1994)
127. J. Wirstam, Chiral symmetry in two-color QCD at finite temperature. Phys. Rev. **D62**, 045012 (2000)
128. J. Wirstam, J.T. Lenaghan, K. Splittorff, Melting the diquark condensate in two-color QCD: A renormalization group analysis. Phys. Rev. **D67**, 034021 (2003)

129. J. Polonyi, K. Szlachanyi, Phase transition from strong coupling expansion. Phys. Lett. **B110**, 395–398 (1982)
130. E. Dagotto, F. Karsch, A. Moreo, The strong coupling limit of SU(2) QCD at finite baryon density. Phys. Lett. **B169**, 421 (1986)
131. E. Dagotto, A. Moreo, U. Wolff, Study of lattice SU(N) QCD at finite baryon density. Phys. Rev. Lett. **57**, 1292 (1986)
132. E. Dagotto, A. Moreo, U. Wolff, Lattice SU(N) QCD at finite temperature and density in the strong coupling limit. Phys. Lett. **B186**, 395 (1987)
133. Y. Nishida, K. Fukushima, T. Hatsuda, Thermodynamics of strong coupling 2-color QCD with chiral and diquark condensates. Phys. Rept. **398**, 281–300 (2004)
134. S. Chandrasekharan, F.-J. Jiang, Phase-diagram of two-color lattice QCD in the chiral limit. Phys. Rev. **D74**, 014506 (2006)
135. S. Chandrasekharan, Anomalous superfluidity in 2+1 dimensional two-color lattice QCD. Phys. Rev. Lett. **97**, 182001 (2006)
136. K. Fukushima, Characteristics of the eigenvalue distribution of the Dirac operator in dense two-color QCD. JHEP **07**, 083 (2008)
137. B. Vanderheyden, A.D. Jackson, Random matrix study of the phase structure of QCD with two colors. Phys. Rev. **D64**, 074016 (2001)
138. B. Klein, D. Toublan, J.J.M. Verbaarschot, Diquark and pion condensation in random matrix models for two-color QCD. Phys. Rev. **D72**, 015007 (2005)
139. Y. Shinno, H. Yoneyama, Analytic Continuation in Two-color Finite Density QCD and Chiral Random Matrix Model, arXiv:0903.0922
140. T. Kanazawa, T. Wettig, N. Yamamoto, Chiral random matrix theory for two-color QCD at high density. Phys. Rev. **D81**, 081701 (2010)
141. W.E. Pauli, On the conservation of the lepton charge. Nuovo Cimento **6**, 205 (1957)
142. F. Gürsey, Relation of charge independence and baryon conservation to Pauli's transformation. Nuovo Cimento **7**, 411 (1958)
143. T. Kanazawa, T. Wettig, N. Yamamoto, Singular values of the Dirac operator in dense QCD-like theories. JHEP **12**, 007 (2011)
144. M.E. Peskin, The alignment of the vacuum in theories of technicolor. Nucl. Phys. **B175**, 197–233 (1980)
145. J. Preskill, Subgroup alignment in hypercolor theories. Nucl. Phys. **B177**, 21–59 (1981)
146. D.A. Kosower, Symmetry breaking patterns in pseudoreal and real gauge theories. Phys. Lett. **B144**, 215–216 (1984)
147. C. Vafa, E. Witten, Restrictions on symmetry breaking in vector-like gauge theories. Nucl. Phys. **B234**, 173 (1984)
148. S. Aoki, A. Gocksch, Spontaneous breaking of flavor symmetry and parity in lattice QCD with Wilson fermions. Phys. Rev. **D45**, 3845–3853 (1992)
149. S.R. Sharpe, R.L. Singleton Jr, Spontaneous flavor and parity breaking with Wilson fermions. Phys. Rev. **D58**, 074501 (1998)
150. T.D. Cohen, Spontaneous parity violation in QCD at finite temperature: On the inapplicability of the Vafa-Witten theorem. Phys. Rev. **D64**, 047704 (2001)
151. H. Leutwyler, A.V. Smilga, Spectrum of Dirac operator and role of winding number in QCD. Phys. Rev. **D46**, 5607–5632 (1992)
152. J.J.M. Verbaarschot, The Spectrum of the QCD Dirac operator and chiral random matrix theory: The Three-fold way. Phys. Rev. Lett. **72**, 2531–2533 (1994)
153. S. Scherer, Introduction to chiral perturbation theory. Adv. Nucl. Phys. **27**, 277 (2003)
154. D.B. Kaplan, Five lectures on effective field theory, nucl-th/0510023
155. B. Kubis, An introduction to chiral perturbation theory, hep-ph/0703274
156. M. Golterman, Applications of chiral perturbation theory to lattice QCD, arXiv:0912.4042
157. S. Weinberg, The Quantum Theory of Fields. Vol. 2: Modern Applications, (Cambridge, UK 1996) p. 489
158. S. Hashimoto, Spontaneous chiral symmetry breaking on the lattice, PoS CD09 004 (2009)

159. G. 't Hooft, Naturalness, chiral symmetry, and spontaneous chiral symmetry breaking. NATO Adv. Study Inst. Ser. B Phys. 59, 135 (1980)
160. S.R. Coleman, E. Witten, Chiral symmetry breakdown in large N chromodynamics. Phys. Rev. Lett. **45**, 100 (1980)
161. J. Polchinski, Effective field theory and the Fermi surface, hep-th/9210046
162. D. B. Kaplan, Effective field theories, nucl-th/9506035
163. A.V. Manohar, Effective field theories, hep-ph/9606222
164. C.P. Burgess, Goldstone and pseudo-Goldstone bosons in nuclear, particle and condensed-matter physics. Phys. Rept. **330**, 193–261 (2000)
165. T. Brauner, Spontaneous symmetry breaking and Nambu-Goldstone Bosons in quantum Many-Body systems. Symmetry **2**, 609–657 (2010)
166. S. Weinberg, Phenomenological lagrangians. Physica **A96**, 327 (1979)
167. J. Gasser, H. Leutwyler, On the low-energy structure of QCD. Phys. Lett. **B125**, 321 (1983)
168. J. Gasser, H. Leutwyler, Chiral perturbation theory to one loop. Ann. Phys. **158**, 142 (1984)
169. H. Kluberg-Stern, A. Morel, B. Petersson, Spectrum of lattice gauge theories with fermions from a 1/D expansion at strong coupling. Nucl. Phys. **B215**, 527 (1983)
170. J.B. Kogut, H. Matsuoka, M. Stone, H.W. Wyld, S.H. Shenker, J. Shigemitsu, D.K. Sinclair, Chiral symmetry restoration in baryon rich environments. Nucl. Phys. **B225**, 93 (1983)
171. S.J. Hands, M. Teper, On the value and origin of the chiral condensate in quenched SU(2) lattice gauge theory. Nucl. Phys. **B347**, 819–853 (1990)
172. T. Banks, A. Casher, Chiral symmetry breaking in confining theories. Nucl. Phys. **B169**, 103 (1980)
173. T. Schafer, E.V. Shuryak, Instantons in QCD. Rev. Mod. Phys. **70**, 323–426 (1998)
174. D. Diakonov, Instantons at work. Prog. Part. Nucl. Phys. **51**, 173–222 (2003)
175. A.V. Smilga, J. Stern, On the spectral density of Euclidean Dirac operator in QCD. Phys. Lett. **B318**, 531–536 (1993)
176. D. Toublan, J.J.M. Verbaarschot, Dirac spectrum in adjoint QCD, hep-th/0008086
177. J.C. Osborn, D. Toublan, J.J.M. Verbaarschot, From chiral random matrix theory to chiral perturbation theory. Nucl. Phys. **B540**, 317–344 (1999)
178. P.H. Damgaard, J.C. Osborn, D. Toublan, J.J.M. Verbaarschot, The microscopic spectral density of the QCD Dirac operator. Nucl. Phys. **B547**, 305–328 (1999)
179. D. Toublan, J.J.M. Verbaarschot, The spectral density of the QCD Dirac operator and patterns of chiral symmetry breaking. Nucl. Phys. **B560**, 259–282 (1999)
180. C.W. Bernard, M.F.L. Golterman, Partially quenched gauge theories and an application to staggered fermions. Phys. Rev. **D49**, 486–494 (1994)
181. S.R. Sharpe, N. Shoresh, Partially quenched chiral perturbation theory without $\Phi_0$. Phys. Rev. **D64**, 114510 (2001)
182. J.J.M. Verbaarschot, The supersymmetric method in random matrix theory and applications to QCD. AIP Conf. Proc. **744**, 277–362 (2005)
183. J. Gasser, H. Leutwyler, Thermodynamics of chiral symmetry. Phys. Lett. **B188**, 477 (1987)
184. H. Leutwyler, Energy levels of light quarks confined to a box. Phys. Lett. **B189**, 197 (1987)
185. P.H. Damgaard, H. Fukaya, The chiral condensate in a finite volume. JHEP **01**, 052 (2009)
186. P.H. Damgaard, S.M. Nishigaki, Universal spectral correlators and massive Dirac operators. Nucl. Phys. **B518**, 495–512 (1998)
187. J.J.M. Verbaarschot, Spectral sum rules and Selberg's integral formula. Phys. Lett. **B329**, 351–357 (1994)
188. E.P. Wigner, On a class of analytic functions from the quantum theory of collisions. Ann. Math. **53**, 36 (1951)
189. T. Guhr, A. Müller-Groeling, H.A. Weidenmüller, Random matrix theories in quantum physics: Common concepts. Phys. Rept. **299**, 189–425 (1998)
190. P.L. Ferrari, R. Frings, On the partial connection between random matrices and interacting particle systems, arXiv:1006.3946
191. P. Di Francesco, P.H. Ginsparg, J. Zinn-Justin, 2-D Gravity and random matrices. Phys. Rept. **254**, 1–133 (1995)

192. J.B. Conrey, L-functions and random matrices, math/0005300
193. J.P. Bouchaud, M. Potters, Financial applications of random matrix theory: a short review, arXiv:0910.1205
194. A. M. Tulino and S. Verdú, Random Matrix Theory and Wireless Communications. now Publishers Inc., Hanover, 2004.
195. M.L. Mehta, *Random Matrices*, 3rd edn. (Academic Press, Amsterdam, 2004)
196. P.J. Forrester, *Log-Gases and Random Matrices* (Princeton University Press, Princeton, 2010)
197. O. Bohigas, M.J. Giannoni, C. Schmit, Characterization of chaotic quantum spectra and universality of level fluctuation laws. Phys. Rev. Lett. **52**, 1–4 (1984)
198. S. Muller, S. Heusler, A. Altland, P. Braun, F. Haake, Periodic-orbit theory of universal level correlations in quantum chaos. New J. Phys. **11**, 103025 (2009)
199. E.V. Shuryak, J.J.M. Verbaarschot, Random matrix theory and spectral sum rules for the Dirac operator in QCD. Nucl. Phys. **A560**, 306–320 (1993)
200. J.J.M. Verbaarschot, I. Zahed, Spectral density of the QCD Dirac operator near zero virtuality. Phys. Rev. Lett. **70**, 3852–3855 (1993)
201. J.J.M. Verbaarschot, Universal behavior in Dirac spectra, hep-th/9710114.
202. J.J.M. Verbaarschot, T. Wettig, Random matrix theory and chiral symmetry in QCD. Ann. Rev. Nucl. Part. Sci. **50**, 343–410 (2000)
203. J.J.M. Verbaarschot, QCD, chiral random matrix theory and integrability, hep-th/0502029
204. G. Akemann, Matrix models and QCD with chemical potential. Int. J. Mod. Phys. **A22**, 1077–1122 (2007)
205. J.J.M. Verbaarschot, in Handbook Article on Applications of Random Matrix Theory to QCD, ed. by G. Akemann, J. Baik, P. Di Francesco. The Oxford Handbook of Random Matrix Theory, (Oxford University Press, Oxford, 2011)
206. F.J. Dyson, The three-fold way. Algebraic structure of symmetry groups and ensembles in quantum mechanics. J. Math. Phys. **3**, 1199 (1962)
207. M.R. Zirnbauer, Riemannian symmetric superspaces and their origin in random-matrix theory. J. Math. Phys. **37**, 4986–5018 (1996)
208. M.R. Zirnbauer, Symmetry Classes, arXiv:1001.0722
209. J.J.M. Verbaarschot, I. Zahed, Random matrix theory and QCD in three-dimensions. Phys. Rev. Lett. **73**, 2288–2291 (1994)
210. R.J. Szabo, Microscopic spectrum of the QCD Dirac operator in three dimensions. Nucl. Phys. **B598**, 309–347 (2001)
211. F. Basile, G. Akemann, Equivalence of QCD in the epsilon-regime and chiral random matrix theory with or without chemical potential. JHEP **12**, 043 (2007)
212. G. Akemann, P.H. Damgaard, U. Magnea, S. Nishigaki, Universality of random matrices in the microscopic limit and the Dirac operator spectrum. Nucl. Phys. **B487**, 721–738 (1997)
213. R. Pullirsch, K. Rabitsch, T. Wettig, H. Markum, Evidence for quantum chaos in the plasma phase of QCD. Phys. Lett. **B427**, 119–124 (1998)
214. A.M. Halasz, A.D. Jackson, R.E. Shrock, M.A. Stephanov, J.J.M. Verbaarschot, On the phase diagram of QCD. Phys. Rev. **D58**, 096007 (1998)
215. B. Klein, D. Toublan, J.J.M. Verbaarschot, The QCD phase diagram at nonzero temperature, baryon and isospin chemical potentials in random matrix theory. Phys. Rev. **D68**, 014009 (2003)
216. P.H. Damgaard, K. Splittorff, Partially quenched chiral perturbation theory and the replica method. Phys. Rev. **D62**, 054509 (2000)
217. P.H. Damgaard, M.C. Diamantini, P. Hernandez, K. Jansen, Finite-size scaling of meson propagators. Nucl. Phys. **B629**, 445–478 (2002)
218. K. Splittorff, J.J.M. Verbaarschot, Replica limit of the Toda lattice equation. Phys. Rev. Lett. **90**, 041601 (2003)
219. K. Splittorff, J.J.M. Verbaarschot, Factorization of correlation functions and the replica limit of the Toda lattice equation. Nucl. Phys. **B683**, 467–507 (2004)
220. M. Adler, E. Horozov, P. van Moerbeke, The pfaff lattice and skew-orthogonal polynomials. Intern. Math. Res. Notices **11**, 569–588 (1999)

221. M. Adler, P. van Moerbeke, Symmetric random matrices and the pfaff lattice, solv-int/9903009
222. M. Adler, P. van Moerbeke, The pfaff lattice, matrix integrals and a map from toda to pfaff, solv-int/9912008
223. M. Adler, P. van Moerbeke, Toda versus pfaff lattice and related polynomials. Duke Math. J. **112**, 1–58 (2002)
224. M. Kieburg, T. Guhr, Derivation of determinantal structures for random matrix ensembles in a new way. J. Phys. **A43**, 075201 (2010)
225. M. Kieburg, T. Guhr, A new approach to derive Pfaffian structures for random matrix ensembles. J. Phys. **A43**, 135204 (2010)
226. J. Ginibre, Statistical ensembles of complex, quaternion and real matrices. J. Math. Phys. **6**, 440–449 (1965)
227. R. Grobe, F. Haake, H.-J. Sommers, Quantum distinction of regular and chaotic dissipative motion. Phys. Rev. Lett. **61**, 1899–1902 (1988)
228. L.E. Reichl, C. Zhong-Ying, M. Millonas, Stochastic manifestation of chaos in a Fokker-Planck equation. Phys. Rev. Lett. **63**, 2013 (1989)
229. H.J. Sommers, A. Crisanti, H. Sompolinsky, Y. Stein, The spectrum of large random asymmetric matrices. Phys. Rev. Lett. **60**, 1895 (1988)
230. H. Sompolinsky, A. Crisanti, Chaos in random neural networks. Phys. Rev. Lett. **61**, 259 (1988)
231. Y.V. Fyodorov, H.J. Sommers, Statistics of resonance poles, phase shifts and time delays in quantum chaotic scattering for systems with broken time reversal invariance. J. Math. Phys. **38**, 1918 (1997)
232. N. Hatano, D.R. Nelson, Localization transitions in non-hermitian quantum mechanics. Phys. Rev. Lett. **77**, 570–573 (1996)
233. N. Hatano, D.R. Nelson, Vortex pinning and non-hermitian quantum mechanics. Phys. Rev. B **56**, 8651–8673 (1997)
234. D. Bernard, A. LeClair, A classification of non-hermitian random matrices, cond-mat/0110649
235. U. Magnea, Random matrices beyond the Cartan classification. J. Phys. **A41**, 045203 (2008)
236. M.A. Stephanov, Random matrix model of QCD at finite density and the nature of the quenched limit. Phys. Rev. Lett. **76**, 4472–4475 (1996)
237. J.C. Osborn, Universal results from an alternate random matrix model for QCD with a baryon chemical potential. Phys. Rev. Lett. **93**, 222001 (2004)
238. G. Akemann, J.C. Osborn, K. Splittorff, J.J.M. Verbaarschot, Unquenched QCD Dirac operator spectra at nonzero baryon chemical potential. Nucl. Phys. **B712**, 287–324 (2005)
239. J.C. Osborn, K. Splittorff, J.J.M. Verbaarschot, Chiral symmetry breaking and the Dirac spectrum at nonzero chemical potential. Phys. Rev. Lett. **94**, 202001 (2005)
240. J.C. Osborn, K. Splittorff, J.J.M. Verbaarschot, Chiral condensate at nonzero chemical potential in the microscopic limit of QCD. Phys. Rev. **D78**, 065029 (2008)
241. J.C. Osborn, K. Splittorff, J.J.M. Verbaarschot, Phase diagram of the dirac spectrum at nonzero chemical potential. Phys. Rev. **D78**, 105006 (2008)
242. J.C. Osborn, Partially quenched QCD with a chemical potential, PoS LAT2006 142 (2006).
243. D. Toublan, J.J.M. Verbaarschot, Effective low energy theories and QCD Dirac spectra. Int. J. Mod. Phys. **B15**, 1404–1415 (2001)
244. J.C. Osborn, T. Wettig, Dirac eigenvalue correlations in quenched QCD at finite density, PoS LAT2005 200 (2006).
245. G. Akemann, The complex Laguerre symplectic ensemble of non-Hermitian matrices. Nucl. Phys. **B730**, 253–299 (2005)
246. G. Akemann, F. Basile, Massive partition functions and complex eigenvalue correlations in Matrix Models with symplectic symmetry. Nucl. Phys. **B766**, 150–177 (2007)
247. G. Akemann, E. Bittner, Unquenched complex Dirac spectra at nonzero chemical potential: Two-colour QCD lattice data versus matrix model. Phys. Rev. Lett. **96**, 222002 (2006)
248. Y.V. Fyodorov, B.A. Khoruzhenko, H.-J. Sommers, Almost-hermitian random matrices: eigenvalue density in the complex plane. Phys. Lett. **A226**, 46–52 (1997)

249. Y.V. Fyodorov, B.A. Khoruzhenko, H.-J. Sommers, Almost hermitian random matrices: crossover from wigner- dyson to ginibre eigenvalue statistics. Phys. Rev. Lett. **79**, 557–560 (1997)
250. Y.V. Fyodorov, B. Khoruzhenko, H.J. Sommers, Universality in the random matrix spectra in the regime of weak non-Hermiticity. Ann. Inst. Henri Poincare (Physique Theorique) **68**, 449–489 (1998)

# Chapter 3
# Dirac Operator in Dense QCD

This and the next chapter constitute the main part of this thesis. In this chapter, we will show that exact analysis of the Dirac operator is possible in the high density BCS phase of QCD for two and three colors, where the symmetries are spontaneously broken by the diquark condensate $\langle qq \rangle$, rather than by the conventional chiral condensate $\langle \bar{q}q \rangle$. In Sects. 3.1 and 3.2 we define and analyze the $\varepsilon$-regime at high density. We will show a novel link between the spectrum of low-lying Dirac eigenvalues and the BCS pairing gap. In Sects. 3.3 and 3.4 we introduce and solve ChRMT for two-color QCD at nonzero chemical potential. We will analyze in detail the dependence of the microscopic Dirac spectrum on the quark masses and the chemical potential. Finally in Sect. 3.5 we analyze the sign problem based on the exact results in the previous section.

## 3.1 Dense QCD in a Finite Volume

In this section we demonstrate exact analysis for the Dirac operator in $N_c = N_f = 3$ QCD at asymptotically large quark chemical potential $\mu$ in a large but finite volume [1]. This is a direct extension of the work by Leutwyler and Smilga at $\mu = 0$ [2] reviewed in Chap. 2 Sect. 2.3.2. The exact results in two limits ($\mu = 0$ and $\mu \to \infty$) impose strong constraints on the possible Dirac spectra and provide important insights to the properties of QCD at finite $\mu$.

As reviewed in Chap. 2 Sect. 2.1.2, the ground state of $N_c = N_f = 3$ QCD at high density is characterized by the diquark pairing $\langle (q_L)^j_b C (q_L)^k_c \rangle \sim \varepsilon_{abc}\varepsilon_{ijk}[d_L^\dagger]_{ai}$ and $\langle (q_R)^j_b C (q_R)^k_c \rangle \sim \varepsilon_{abc}\varepsilon_{ijk}[d_R^\dagger]_{ai}$ where $i, j, k$ $(a, b, c)$ are the flavor (color) indices, and $C$ is the charge conjugation matrix. The remarkable feature here is that chiral symmetry is dynamically broken as a result of the color-flavor locking (CFL) even though $\langle \bar{q}q \rangle = 0$. The symmetry breaking pattern of the CFL phase at asymptotically large $\mu$ is

T. Kanazawa, *Dirac Spectra in Dense QCD*, Springer Theses,
DOI: 10.1007/978-4-431-54165-3_3, © Springer Japan 2013

$$\mathrm{SU}(3)_C \times \mathrm{U}(1)_A \times \mathrm{U}(1)_B \times \mathrm{SU}(3)_R \times \mathrm{SU}(3)_L \to (\mathbb{Z}_2)_R \times (\mathbb{Z}_2)_L \times \mathrm{SU}(3)_{C+R+L}, \quad (3.1)$$

where compared to (2.5) the $\mathrm{U}(1)_A$ symmetry is added here to incorporate the anomaly suppression at asymptotically large $\mu$ (see Chap. 2 Sect. 2.1.2). The $(\mathbb{Z}_2)_R \times (\mathbb{Z}_2)_L$ symmetry reflects the fact that we can change the sign of the left-handed or right-handed quark fields independently. As a result, we have 8+1+1 Nambu–Goldstone (NG) modes associated with the breaking of chiral symmetry, $\mathrm{U}(1)_A$ and $\mathrm{U}(1)_B$ symmetries, which we will refer to as $\pi^a$, $\eta'$ and $H$, respectively. In the following, we will not consider $H$ since its dynamics decouples.

In the CFL phase, all gluons as well as quarks acquire masses $\sim\Delta \gg \Lambda_{\mathrm{QCD}}$ (Chap. 2 Eq. (2.2)), hence it is viable to construct an effective theory for NG bosons in the CFL phase which is valid at energies well below the BCS gap $\Delta$. Let us introduce

$$\tilde{\Sigma} = \exp(i\pi^a T^a/f_\pi) \in \mathrm{SU}(3), \qquad V = \exp(2i\eta'/(\sqrt{6}f_{\eta'})) \in \mathrm{U}(1) \quad (3.2)$$

with $f_\pi$ ($f_{\eta'}$) the pion ($\eta'$) decay constant. Now we can constrain possible mass terms based on a symmetry principle. Owing to the $(\mathbb{Z}_2)_R \times (\mathbb{Z}_2)_L$ symmetry the leading mass term in the effective theory appears at $\mathcal{O}(M^2)$.[1] Terms allowed by symmetries are

$$V(\mathrm{tr}\, M\tilde{\Sigma}^\dagger)^2 + \mathrm{h.c.}, \quad V\,\mathrm{tr}(M\tilde{\Sigma}^\dagger)^2 + \mathrm{h.c.}, \quad \text{and} \quad \mathrm{tr}[M\tilde{\Sigma}^\dagger]\,\mathrm{tr}[M^\dagger\tilde{\Sigma}]. \quad (3.3)$$

This is the best one can learn from symmetries alone. However, for sufficiently large density we can use weak coupling methods to determine the coefficients in the effective theory. The leading-order result thereby obtained reads [3–6]

$$\begin{aligned}\mathcal{L}_{\mathrm{EFT}} = &\frac{f_\pi^2}{4}\mathrm{tr}[\nabla_0\tilde{\Sigma}\nabla_0\tilde{\Sigma}^\dagger - v_\pi^2\partial_i\tilde{\Sigma}\partial_i\tilde{\Sigma}^\dagger] \\ &+\frac{3f_{\eta'}^2}{4}\left(\partial_0 V\partial_0 V^* - v_{\eta'}^2\partial_i V\partial_i V^*\right) \\ &+\frac{3\Delta^2}{4\pi^2}\{V(\mathrm{tr}\, M\tilde{\Sigma}^\dagger)^2 - V\,\mathrm{tr}(M\tilde{\Sigma}^\dagger)^2 + \mathrm{h.c.}\},\end{aligned} \quad (3.4)$$

where $v_\pi$ ($v_{\eta'}$) is the pion ($\eta'$) velocity, and the covariant derivative including the effective chemical potential (Bedaque-Schäfer term [7]) is given by

$$\nabla_0\tilde{\Sigma} = \partial_0\tilde{\Sigma} + i\left(\frac{MM^\dagger}{2p_F}\right)\tilde{\Sigma} - i\tilde{\Sigma}\left(\frac{M^\dagger M}{2p_F}\right) \quad (3.5)$$

[1] Precisely speaking, this symmetry is slightly violated by nonzero chiral condensate generated by instanton effects. At asymptotically large $\mu$ it is negligibly small and one can neglect it in the following.

with the Fermi momentum $p_F$. The quantities $f_\pi$ and $f_{\eta'}$ can be computed perturbatively at $\mu \gg \Lambda_{\rm QCD}$ as [3–6]

$$\frac{f_\pi^2}{p_F^2} = \frac{21 - 8\log 2}{36\pi^2} \quad \text{and} \quad \frac{f_{\eta'}^2}{p_F^2} = \frac{3}{8\pi^2}. \tag{3.6}$$

The coefficient $\Delta^2$ (the square of the BCS gap) of the mass term Eq. (3.4) is derived from a mass term in the microscopic theory; see the Appendix A.2 for a brief summary of the procedure. While the Lagrangian shown above is the leading order term, it is known that terms in the effective Lagrangian have a generic form [8]

$$\mathcal{L} \sim f_\pi^2 \Delta^2 \left(\frac{\partial_0 - iMM^\dagger/(2p_F)}{\Delta}\right)^n \left(\frac{\partial}{\Delta}\right)^m \left(\frac{MM}{p_F^2}\right)^p. \tag{3.7}$$

We proceed to define the microscopic domain (or the $\varepsilon$-regime) of the CFL phase as follows. Consider QCD on a four-dimensional torus of size $V_4 = L \times L \times L \times \beta$ with $\beta = 1/T \sim L$. The microscopic domain of QCD at $\mu = 0$ is specified by $\dfrac{1}{\Lambda_{\rm QCD}} \ll L \ll \dfrac{1}{m_\Pi}$ (cf. Chap. 2 Eq. (2.65)), where $m_\Pi$ denotes the mass of pions at $\mu = 0$ and $\Lambda_{\rm QCD}$ characterizes the mass scale of non-NG modes (e.g., the $\rho$ meson mass $m_\rho$) [2]. Analogously, the microscopic domain of the CFL phase can be defined as

$$\frac{1}{\Delta} \ll L \ll \frac{1}{m_{\pi,\eta'}}, \tag{3.8}$$

where $m_{\pi,\eta'}$ is the mass of NG modes associated with the symmetry breaking at high $\mu$: (3.1). The first condition in (3.8) follows by comparing the contribution to $Z_{\rm EFT}$ of NG modes, $e^{-m_{\pi,\eta'}L}$, to that of the other heavier particles, $e^{-\Delta L}$. This condition vindicates the use of the CFL effective theory. The second condition in (3.8) means that the Compton wavelength of the pions is much larger than the linear size of the box, so that the CFL effective Lagrangian can be truncated to its zero momentum sector. This condition is automatically satisfied at sufficiently high $\mu$, even with a finite quark mass, since $\Delta/\mu \ll 1$ in $m_{\pi,\eta'} \sim \Delta m/\mu$ (see (3.4). Note also that, in the domain (3.8), the temperature $T$ is low enough for the CFL phase to be realized: $T \ll T_c \sim \Delta$ with $T_c$ being the critical temperature of the color superconductivity.

The QCD partition function involves a sum over the integer topological charge $\nu$. At $\mu \gg \Lambda_{\rm QCD}$, however, the topological susceptibility is highly suppressed as $\langle \nu^2 \rangle \propto (\Lambda_{\rm QCD}/\mu)^8$ [9, 10], thus we can focus on the topological sector $\nu = 0$ alone.

Therefore the partition function for the CFL phase in the $\varepsilon$-regime is given by

**Table 3.1** Summary of results at $\mu \gg \Lambda_{\text{QCD}}$ [1] compared with those at $\mu = 0$ [2]: order parameters, the corresponding source terms, and the linear spacing of eigenvalues $\delta\lambda_n$, respectively

| | Order parameter | Source | $\delta\lambda_n$ |
|---|---|---|---|
| $\mu \gg \Lambda_{\text{QCD}}$ | $\Delta^2 \propto \lvert d_L^\dagger d_R \rvert/\mu^4$ | $m^2$ | $\propto 1/\sqrt{V_4}$ |
| $\mu = 0$ | $\langle \bar{q}q \rangle$ | $m$ | $\propto 1/V_4$ |

$$Z_{\text{EFT}}(M) = \int_{\text{U}(3)} [d\Sigma] \exp\left( \frac{3}{4\pi^2} V_4 \Delta^2 \left\{ (\text{tr}\, M\Sigma^\dagger)^2 - \text{tr}(M\Sigma^\dagger)^2 \right\} \det\Sigma + \text{c.c.} \right), \tag{3.9}$$

where $\Sigma \equiv \tilde{\Sigma} V \in \text{U}(3)$. We have neglected the effect of the effective chemical potential, (3.5). (If one includes it, (3.9) would be expanded in terms of not only $(V_4\Delta^2)^2 \mathcal{O}(M^4)$ but also $V_4 \mathcal{O}(M^4)$. In the domain $\Delta^{-1} \ll L$, however, the latter is negligible). We have also neglected a term of the form $\text{tr}[MM^\dagger]$ since it does not contain the pion fields, but there is a subtlety on this; see the Appendix A.3 for a careful discussion. $Z_{\text{EFT}}$ is normalized to be 1 in the chiral limit. It is worth noting that $M^2$ acts as a source for $\Delta^2$.

Owing to the property of $\Sigma \in \text{U}(3)$, $\left\{ (\text{tr}\, M\Sigma^\dagger)^2 - \text{tr}(M\Sigma^\dagger)^2 \right\} \det\Sigma = 2 \det M\, \text{tr}[M^{-1}\Sigma]$, (3.9) turns out to be of the same form as the conventional $\varepsilon$-regime partition function at $\mu = 0$ and $\nu = 0$, (2.69), but with $M$ replaced by $(\det M) M^{-1}$ and the chiral condensate replaced by $\Delta^2$. This is a novel correspondence between the color superconducting phase and the hadronic phase, and may have relevance to the idea of their continuity [11]. In Table 3.1, main results at $\mu \gg \Lambda_{\text{QCD}}$ are compared with the results at $\mu = 0$ in [2].

Exploiting this correspondence and the exact result at $\mu = 0$ in [12], the integral (3.9) can be computed analytically. In particular, in the flavor symmetric case $M = m\mathbb{1}$, we have

$$Z_{\text{EFT}}(m) = \det_{0 \le i, j \le 2} \left[ I_{j-i}(x) \right], \tag{3.10}$$

where $I_\nu(x)$ is the modified Bessel function and $x = 3V_4\Delta^2 m^2/\pi^2$.

For a generic mass matrix, (3.9) may be expanded in powers of $M$ followed by the group integral over U(3), resulting in

$$Z_{\text{EFT}}(M) = 1 + \frac{3}{8\pi^4} (V_4\Delta^2)^2 \left\{ (\text{tr}\, M^\dagger M)^2 - \text{tr}(M^\dagger M)^2 \right\} + \mathcal{O}(M^8). \tag{3.11}$$

Next, we turn to the microscopic expression of the QCD partition function. Let $\{i\lambda_n\}_n$ $(\lambda_n \in \mathbb{C})$ denote the eigenvalues of the Euclidean Dirac operator $\mathcal{D}(\mu)$. The chiral symmetry $\{\mathcal{D}(\mu), \gamma_5\} = 0$ ensures that if $i\lambda_n$ is an eigenvalue of $\mathcal{D}(\mu)$, so is $-i\lambda_n$. The QCD partition function with $\nu = 0$ can be expressed as

$$Z_{\text{QCD}}(M) = \left\langle \det\left( \mathcal{D}(\mu) + \frac{1-\gamma_5}{2}M + \frac{1+\gamma_5}{2}M^\dagger \right) \right\rangle_{N_f=0} \tag{3.12}$$

$$= \left\langle \prod_{\text{Re}(\lambda_n)>0} \det(\lambda_n^2 + M^\dagger M) \right\rangle_{N_f=0}. \tag{3.13}$$

Normalizing $Z_{\text{QCD}}$ to 1 in the chiral limit, we find

$$Z_{\text{QCD}}(M) = \left\langle\!\!\left\langle \prod_{\text{Re}(\lambda_n)>0} \det\left(1 + \frac{M^\dagger M}{\lambda_n^2}\right) \right\rangle\!\!\right\rangle, \tag{3.14}$$

where $\langle\!\langle \mathcal{O} \rangle\!\rangle = \int [dA]\, \mathcal{O} \mathrm{e}^{-S_g} \Big(\prod_n \lambda_n^2\Big)^3 \Big/ \int [dA] \mathrm{e}^{-S_g} \Big(\prod_n \lambda_n^2\Big)^3$ is the average of $\mathcal{O}$ over all gauge configurations in the chiral limit, with $S_g$ being the action of the gluon field.

Using the relation $\det[1+\varepsilon] = 1 + \operatorname{tr}\varepsilon + \frac{1}{2}\Big\{(\operatorname{tr}\varepsilon)^2 - \operatorname{tr}[\varepsilon^2]\Big\} + \mathcal{O}(\varepsilon^3)$, one can expand the QCD partition function (3.14) in powers of the quark mass matrix $M$. Then one obtains the spectral sum rules for the Dirac eigenvalues $i\lambda_n$ by matching this expansion against (3.11). After a rescaling $z_n \equiv \sqrt{V_4}\Delta\lambda_n$ we find

$$\left\langle\!\!\left\langle \sum_n{}' \frac{1}{z_n^4} \right\rangle\!\!\right\rangle = \left\langle\!\!\left\langle \left(\sum_n{}' \frac{1}{z_n^2}\right)^2 \right\rangle\!\!\right\rangle = \frac{3}{4\pi^4}, \tag{3.15}$$

$$\left\langle\!\!\left\langle \sum_n{}' \frac{1}{z_n^2} \right\rangle\!\!\right\rangle = \left\langle\!\!\left\langle \sum_n{}' \frac{1}{z_n^6} \right\rangle\!\!\right\rangle = \left\langle\!\!\left\langle \left(\sum_n{}' \frac{1}{z_n^2}\right)^3 \right\rangle\!\!\right\rangle$$
$$= \left\langle\!\!\left\langle \left(\sum_n{}' \frac{1}{z_n^2}\right)\left(\sum_n{}' \frac{1}{z_n^4}\right) \right\rangle\!\!\right\rangle = 0, \quad \text{etc.}, \tag{3.16}$$

where the summation $\sum'$ is over $z_n$ satisfying $\text{Re}(z_n) > 0$ and $|z_n| \lesssim \sqrt{V_4}\Delta^2$ ($|\lambda_n| \lesssim \Delta$). These relations are highly nontrivial, since the averaged sums of inverse powers of *complex* $\lambda_n$ take *real* values. The vanishing of the spectral sum rules in (3.16) is physically rooted in the $(\mathbb{Z}_2)_L \times (\mathbb{Z}_2)_R$ symmetry of the diquark pairing at large $\mu$. The above results should be compared with the spectral sum rules at $\mu = 0$ [2], e.g., (2.74), where the second inverse moment of $\{i\lambda_n\}$ is *nonzero* and the eigenvalues are normalized by the chiral condensate rather than $\Delta^2$.

At first sight, the above results are counter-intuitive in the following sense. Since the $i\lambda_n$ are the eigenvalues of the Dirac operator $\gamma_\nu D_\nu + \gamma_0\mu$ for asymptotically large $\mu$, one naively expects $\lambda_n \sim \mu$, and the inverse moments of the eigenvalues should *decrease* for larger $\mu$. But our exact results imply exactly the opposite[2]!—What we

---

[2] $\Delta$ diverges as $\mu \to \infty$, see Chap. 2 Eq. (2.2).

have overlooked is the existence of the Fermi surface. The typical momentum is comparable to the Fermi momentum, hence $\gamma_\nu D_\nu$ is not necessarily small compared to $\gamma_0\mu$. Delicate cancellations between these two terms lead to the above nontrivial sum rules.

In terms of the spectral density $\rho(\lambda) \equiv \left\langle\!\left\langle \sum_n \delta^2(\lambda - \lambda_n) \right\rangle\!\right\rangle$,[3] the first sum rule (3.15) may be written as

$$\frac{1}{V_4\Delta^2}\int_{\mathbb{C}_+}\frac{d^2z}{z^4}\rho\left(\frac{z}{\sqrt{V_4}\Delta}\right) = \frac{3}{4\pi^4}, \tag{3.17}$$

where $\mathbb{C}_+ = \{z : \text{Re}(z) > 0\}$ and $d^2z \equiv d(\text{Re}z)d(\text{Im}z)$. This implies the existence of the *microscopic limit* of the spectral density,

$$\rho_s(z) \equiv \lim_{V_4\to\infty}\frac{1}{V_4\Delta^2}\rho\left(\frac{z}{\sqrt{V_4}\Delta}\right). \tag{3.18}$$

This is a novel generalization of the microscopic spectral density defined at $\mu = 0$ ($\rightarrow$ 2.85). From (3.18), we find that the microscopic spectral density at large $\mu$ is governed by the color superconducting gap $\Delta$, rather than the chiral condensate. Equation (3.18) also shows that the linear spacing of eigenvalues $\delta\lambda_n$ near the origin on the complex $\lambda$-plane depends on the volume as $\delta\lambda_n \propto 1/\sqrt{V_4}$. Compared to $\delta\lambda_n \propto 1/V_4$ at low $\mu$ [2] and $\delta\lambda_n \propto 1/V_4^{1/4}$ in a free theory, the near-zero Dirac spectrum is deformed substantially owing to the interplay of gauge interactions and the Fermi surface.

Can we expect that the results obtained in the $\varepsilon$-regime at large $\mu$ would be *universal*, i.e., determined solely by symmetries and their breaking pattern? If so, it will be presumably possible to construct a random matrix theory for the CFL phase and use it to compute the microscopic spectral density $\rho_s$ explicitly, as was done at $\mu = 0$ [13, 14]. Alternatively the supersymmetric approach [15, 16], after properly extended to incorporate color superconductivity, might be useful.

However we can envisage a difficulty in this direction. As noted in (3.3), there are three mass terms at $\mathcal{O}(M^2)$ that are consistent with symmetries, and *one cannot determine their relative coefficients based on symmetries alone.* To address this issue one needs detailed information on the UV dynamics of QCD, such as the fact that the one-gluon exchange interaction between quarks is repulsive in the **6** channel but attractive in the $\bar{\mathbf{3}}$ channel. It would therefore be mandatory to input a certain amount of UV information besides the pattern of symmetry breaking, in constructing a suitable random matrix model. Even if it could be done, we still have to invent a method to handle a four-quark condensate within RMT, since a gauge invariant order parameter of the CFL phase at the lowest order in fields is the four-quark condensate. Having a nonzero four-quark condensate, $\langle (qq)^\dagger qq\rangle \neq 0$ in the absence

[3] Definition: $\delta^2(z) \equiv \delta(\text{Re}\, z)\delta(\text{Im}\, z)$.

of the diquark condensate, $\langle qq \rangle = 0$, [4] seems to imply that this phase cannot be captured in the mean field approximation (cf. [17]). ChRMT is, however, a theory of chiral symmetry breaking in the mean field approximation, because it is a zero-dimensional model, hence no "fluctuation". This point seems to be the main difficulty of ChRMT in the application to the CFL phase.

To bypass this difficulty we shall look into another theory in the next section whose ground state is characterized by a color-singlet diquark condensate and has a unique $\mathcal{O}(M^2)$ term. It is two-color QCD.

## 3.2 Dense Two-Color QCD

In this section we will determine the pattern of spontaneous symmetry breaking in two-color QCD at high baryon density and use it to construct the chiral Lagrangian of light NG fields. Then we will define the $\varepsilon$-regime at high density and derive novel spectral sum rules of Dirac eigenvalues. These results can in principle be checked in future lattice simulations, since two-color QCD has no sign problem even at $\mu \neq 0$ (for pairwise degenerate masses). These findings were originally reported in [18, 19]. The description in this section will be terse when overlapping with the previous section.

### *3.2.1 Symmetry Breaking and Chiral Lagrangian*

We reviewed two-color QCD at low density in Chap. 2 Sect. 2.2. The point was that an enlarged flavor symmetry is spontaneously broken. When current quark masses are nonzero, the NG modes acquire small masses. Because some of them have nonzero baryon number, there is a second order superfluid phase transition at $\mu = m_\pi/2$, where $U(1)_B$ is spontaneously broken by the formation of color-singlet diquark condensate $\langle\psi\psi\rangle$. The ground state can be seen as a system of weakly interacting tightly bound molecules (diquarks). The chiral condensate gradually rotates into the diquark condensate as $\mu$ increases further. All of these can be analyzed by the leading-order ChPT as long as $\mu \sim m_\pi \ll \Lambda_{\text{QCD}}$ is satisfied.

It is however nontrivial to understand the physics in the region of intermediate density. When $\mu/m_\pi \gg 1$, ChPT based on the ansatz $SU(2N_f) \rightarrow Sp(2N_f)$ is no longer reliable, since the original symmetry $SU(2N_f)$ is badly broken by nonzero $\mu$ (recall Table 2.1). Consequently, part of pions acquire large masses and become unstable against decaying processes $\pi \rightarrow q + q$ and $\pi \rightarrow q + \bar{q}$ [20]. Therefore we must go beyond the ChPT based on the ansatz $SU(2N_f) \rightarrow Sp(2N_f)$ to study the phase structure of two-color QCD at intermediate densities. So far quite a few studies have been performed based on NJL-type models [20–24], mainly aiming at understanding a qualitative change of the system *from a dilute Bose condensate*

[4] It acquires a nonzero expectation value once the gauge fixing is performed.

**Table 3.2** Qualitative differences in the physics of two-color QCD at large and small $\mu$ (taken from [18]).

| | $\mu \ll \Lambda_{\mathrm{SU(2)}}$ | $\mu \gg \Lambda_{\mathrm{SU(2)}}$ |
|---|---|---|
| Chiral symmetry | Spontaneously broken | Spontaneously broken |
| $m_\pi^2$ | $\propto m$ | $\propto m^2$ |
| Instantons | Abundant | Suppressed |
| $\mathrm{U}(1)_B$ symmetry | Intact | Spontaneously broken |
| $\mathrm{U}(1)_A$ symmetry | Anomalous | Spontaneously broken |
| $\eta'$ | Heavy | Light |
| Condensate | $0 \simeq \langle\psi\psi\rangle \ll \lvert\langle\bar{\psi}\psi\rangle\rvert$ | $0 \simeq \lvert\langle\bar{\psi}\psi\rangle\rvert \ll \langle\psi\psi\rangle$ |
| Mass gap (quarks) | $\sim\Lambda_{\mathrm{SU(2)}}$ | $\sim\Delta \gg \Lambda_{\mathrm{SU(2)}}$ |
| Mass gap (gluons) | $\sim\Lambda_{\mathrm{SU(2)}}$ | $\sim\Lambda'_{\mathrm{SU(2)}} \ll \Lambda_{\mathrm{SU(2)}}$ |

$m$ denotes the (degenerate) current quark mass. $N_f$ is assumed to be even. See text for further explanations

*of tightly bound molecules at low density to a weakly coupled BCS superfluid of overlapping Cooper pairs at high density* [25]. It is a relativistic analogue of a phenomenon called *BEC-BCS crossover* in condensed matter physics [26–28]. A basic observation here is that there is no global order parameter that distinguishes the two regimes. In dense two-color QCD, lattice simulations have been performed vigorously and the results there indicate that indeed the system seems to undergo a qualitative change at intermediate densities [29, 30]. It is even argued that the quarks as well as gluons may experience a deconfinement transition at finite critical chemical potential, but it is still elusive. It should be noted that the existence of gluons is a unique feature of dense two-color QCD compared to usual non-relativistic systems, and further efforts will be necessary from both theoretical and numerical sides to grasp the whole picture of two-color QCD in the transition region. In Table 3.2 main characteristics of the two regimes are summarized (discussed in greater detail below).

At asymptotically high density, however, physics is less elusive. The attractive interaction in the color antisymmetric channel between quarks causes the rearrangement of the Fermi surface, resulting in the condensation of Cooper pairs and the opening of a gap $\Delta$ in the quasiparticle spectrum. Because Pauli principle forces the diquark wave function to be antisymmetric under simultaneous exchanges of color, flavor and spinor indices, we have

$$\langle\psi\psi\rangle \equiv \left\langle \varepsilon_{ab}(\psi_a^T)^i C\gamma_5 (I_{N_f})^{ij}\psi_b^j \right\rangle \neq 0\,, \tag{3.19}$$

where $C$ is the charge-conjugation matrix, $a, b$ $(i, j)$ are color (flavor) indices, respectively, and the antisymmetric flavor matrix $I$ is defined in (2.17). We assumed $s$-wave condensation here, as it makes maximum use of the Fermi surface. At sufficiently high density $\langle\bar{q}q\rangle \simeq 0$ owing to the huge energy gap $(\sim\mu)$ for $\bar{q}$.

Besides the scalar diquark condensate $\psi^T C\gamma_5\psi$ in (3.19), there is in principle also a pseudo-scalar diquark condensate $\psi^T C\psi$. Although the two are not distinguished by a single-gluon exchange interaction, it has been shown [31, 32] that the instanton-induced interaction favors the former. Based on this observation we also neglect the latter condensate in the following.

It is interesting that the diquark condensate (3.19) which is deduced from the BCS mechanism at high density has the same quantum numbers as that in the low-density region (Chap. 2 Sect. 2.2.3). The SU(2) gauge symmetry is not broken in either cases. This continuity of symmetry is the theoretical underpinning of the suspected BEC-BCS crossover [25].

Above we tacitly assumed that $N_f$ is even, because the case of odd $N_f$ requires a careful treatment due to the same reason as in Chap. 2 Sect. 2.2.3. All the analysis below rests on the assumption that $N_f$ is even. We also assume $N_f < 11$ to ensure asymptotic freedom.

By the diquark condensate (3.19) global symmetries of the Lagrangian are spontaneously broken as ($\rightarrow$ Table 2.1)

$$\mathrm{SU}(N_f)_L \times \mathrm{SU}(N_f)_R \times \mathrm{U}(1)_B \times \mathrm{U}(1)_A \rightarrow \mathrm{Sp}(N_f)_L \times \mathrm{Sp}(N_f)_R\,, \tag{3.20}$$

where we treat $\mathrm{U}(1)_A$ as an exact symmetry at high density.

The case of $N_f = 2$ is special since $\mathrm{SU}(2) = \mathrm{Sp}(2)$: there are only two NG bosons ($H$ and $\eta'$) from the breaking of $\mathrm{U}(1)_B \times \mathrm{U}(1)_A$. For $N_f = 4$ and larger, the spectrum contains additional NG bosons $\pi$ whose masses are lifted by a nonzero current quark mass. Their total number is

$$2(N_f^2 - 1) - 2(N_f^2/2 + N_f/2) = N_f^2 - N_f - 2\,. \tag{3.21}$$

We now turn to the construction of the low-energy effective Lagrangian associated with the symmetry breaking pattern (3.20). We introduce dimensionless color-singlet $N_f \times N_f$ matrix fields $D_L$ and $D_R$, where $N_f$ is even and $> 2$ (the case $N_f = 2$ will be considered separately). In terms of the quark fields, they can be expressed as

$$(D_L)^{ij} \sim (\psi_L^T)^i C \psi_L^j\,, \qquad (D_R)^{ij} \sim (\psi_R^T)^i C \psi_R^j\,. \tag{3.22}$$

Under $\mathrm{SU}(N_f)_L \times \mathrm{SU}(N_f)_R \times \mathrm{U}(1)_A \times \mathrm{U}(1)_B$ the quarks transform as

$$\psi_L \rightarrow \mathrm{e}^{i(\alpha+\beta)} g_L \psi_L\,, \qquad \psi_R \rightarrow \mathrm{e}^{-i(\alpha-\beta)} g_R \psi_R\,, \tag{3.23}$$

where $g_i \in \mathrm{SU}(N_f)_i$ ($i = L, R$) and the phases $\alpha$ and $\beta$ are associated with the $\mathrm{U}(1)_A$ and $\mathrm{U}(1)_B$ rotations, respectively. Thus

$$D_L \rightarrow g_L D_L g_L^T \mathrm{e}^{2i(\alpha+\beta)}\,, \qquad D_R \rightarrow g_R D_R g_R^T \mathrm{e}^{-2i(\alpha-\beta)}\,. \tag{3.24}$$

It is convenient to split the fields into U(1) parts $A$, $V$ and the rest,

$$D_L \equiv \Sigma_L A^\dagger V\,, \qquad D_R \equiv \Sigma_R A V\,, \tag{3.25}$$

so that

$$\Sigma_i \to g_i \Sigma_i g_i^T \quad (i = L, R), \qquad A \to A e^{-2i\alpha}, \qquad V \to V e^{2i\beta}. \tag{3.26}$$

The mass term $\bar{\psi}_L M \psi_R + \bar{\psi}_R M^\dagger \psi_L$ is invariant if we treat the mass matrix $M$ as a spurion field and let

$$M \to g_L M g_R^\dagger e^{2i\alpha}. \tag{3.27}$$

The effective Lagrangian composed of $\Sigma_L$, $\Sigma_R$, $A$, $V$, and $M$ should be invariant under (3.26) and (3.27). There is no invariant combination that contains an odd number of factors of $M$, in contrast to (2.32) at low density where the effective Lagrangian contained a term linear in $M$. This difference can be understood as follows. As $N_f$ is even, we can take $g_L = -1$ and $g_R = 1$ (or $g_L = 1$ and $g_R = -1$), under which $M$ transforms as $M \to -M$ while $\Sigma_L$, $\Sigma_R$, $A$, and $V$ remain unchanged. Therefore $M$ must appear in even powers. The point is that the chiral condensate $\langle \bar{\psi}_L \psi_R \rangle$ is negligible at large $\mu$; otherwise a NG field $\tilde{\Sigma} \sim \bar{\psi}_L \psi_R$ would appear that transforms under $g_L = -1$ as $\tilde{\Sigma} \to -\tilde{\Sigma}$ so that odd powers of $M$ could appear in the Lagrangian in combination with $\tilde{\Sigma}$.

At $\mathcal{O}(M^2)$ the real-valued invariant combination is uniquely found to be

$$A^2 \,\mathrm{tr}[M \Sigma_R M^T \Sigma_L^\dagger] + \text{c.c.} \tag{3.28}$$

No mass term appears for $V$ reflecting the fact that $\mathrm{U}(1)_B$ is not violated by a nonzero quark mass. $\Sigma_L$ and $\Sigma_R$, which are decoupled from each other in the chiral limit, are now coupled by $M$. Replacing $\Sigma_{L,R}$ by $\Sigma_{L,R}^T$ does not yield new invariants. This is because the diquark condensate is formed in a flavor antisymmetric channel, i.e., $\Sigma_{L,R}^T = -\Sigma_{L,R}$. In addition, we have $\Sigma\Sigma^\dagger = -\Sigma\Sigma^* = \mathbf{1}_{N_f}$ and $\det \Sigma = 1$ for $\Sigma \in \mathrm{SU}(N_f)/\mathrm{Sp}(N_f)$, which also greatly reduces the number of nontrivial invariants.

We conclude that the effective Lagrangian in the Minkowski spacetime, to lowest order in the derivatives and the quark masses, and assuming $N_f > 2$ and even, is given by[5]

$$\begin{aligned}\mathcal{L} = &\frac{f_H^2}{2}\left\{|\partial_0 V|^2 - v_H^2 |\partial_i V|^2\right\} + \frac{N_f f_{\eta'}^2}{2}\left\{|\partial_0 A|^2 - v_{\eta'}^2 |\partial_i A|^2\right\} \\ &+ \frac{f_\pi^2}{2}\,\mathrm{tr}\left\{|\partial_0 \Sigma_L|^2 - v_\pi^2 |\partial_i \Sigma_L|^2 + (L \leftrightarrow R)\right\} - c\Delta^2\left\{A^2\,\mathrm{tr}[M\Sigma_R M^T \Sigma_L^\dagger] + \text{c.c.}\right\}.\end{aligned} \tag{3.29}$$

Here, $f_H$, $f_{\eta'}$, and $f_\pi$ are the decay constants of $H$, $\eta'$, and $\pi$, respectively, and the $v$'s are the corresponding velocities originating from the absence of Lorentz invariance

---

[5] The Wess-Zumino-Witten term is necessary for completeness of the theory, but it is irrelevant to the ensuing analysis and neglected in the following.

in the medium. The coefficient $c$ is calculable using the method of [6] and found to be

$$c = \frac{3}{4\pi^2} \,. \tag{3.30}$$

The detailed derivation of (3.30) is given in the Appendix A.2. Note that $\mathcal{L}$ has no dependence on the CP-violating $\theta$-parameter. The effective chemical potential of $\mathcal{O}(M^2)$ (the so-called Bedaque–Schäfer term in the CFL phase of three-color QCD, (3.5) is omitted here since it is suppressed by $\sim 1/\mu$ at large $\mu$. Also, its contribution to the finite-volume partition function starts at $\mathcal{O}(M^4)$, which is sufficiently small compared to the leading $\mathcal{O}(M^2)$ term.

From (3.29) we can derive a mass formula for $\pi$ in the flavor-symmetric case $M = m\mathbf{1}_{N_f}$. For this purpose, we define the NG fields $\pi_i^a$ ($i = L, R, a = 1, \dots, (N_f^2 - N_f)/2 - 1$) corresponding to the coset space $\mathrm{SU}(N_f)_i/\mathrm{Sp}(N_f)_i$ as[6]

$$\Sigma_i = U_i I_{N_f} U_i^T = U_i^2 I_{N_f} \,, \qquad U_i = \exp\left(\frac{i\pi_i^a X^a}{2\sqrt{N_f} f_\pi}\right), \qquad [i = L,\ R] \tag{3.31}$$

where $I_{N_f}$ is defined in Chap. 2, Eq. (2.17) while the $X^a$ are the generators of the coset $\mathrm{SU}(N_f)/\mathrm{Sp}(N_f)$ satisfying the normalization $\mathrm{tr}[X^a X^b] = N_f \delta^{ab}$ and the relation $X^a I_{N_f} = I_{N_f}(X^a)^T$. Since the mass term in (3.29) mixes $\pi_L$ and $\pi_R$, it is necessary to diagonalize the mass matrix for $\pi_{L,R}^a$ to obtain the genuine mass eigenvalues by setting

$$\Pi^a = \frac{1}{\sqrt{2}}(\pi_L^a + \pi_R^a) \,, \qquad \tilde{\Pi}^a = \frac{1}{\sqrt{2}}(\pi_L^a - \pi_R^a) \,. \tag{3.32}$$

The resulting mass formula for $\Pi^a$ and $\tilde{\Pi}^a$ reads

$$m_{\Pi^a} = 0 \,, \qquad f_\pi^2 m_{\tilde{\Pi}^a}^2 = 4c\Delta^2 m^2 \,. \tag{3.33}$$

The massless modes $\Pi^a$ correspond to simultaneous rotations of $\pi_L^a$ and $\pi_R^a$ in the same direction, while the massive modes $\tilde{\Pi}^a$ correspond to rotations in the opposite direction (a similar discussion can be found in [33]).[7] For the massive modes $\tilde{\Pi}^a$, (3.33) has exactly the same form as the expression derived in [4], except for the numerical factor of $c$. However, our "pions" for two colors are two-quark ($qq$) states, while those for three colors are four-quark ($\bar{q}\bar{q}qq$) states [4].

Similarly, we can also derive a mass formula for the $\eta'$ boson associated with the spontaneous breaking of $\mathrm{U}(1)_A$. We again assume the flavor-symmetric case $M = m\mathbf{1}_{N_f}$. The $\eta'$ field is defined by

---

[6] For $N_f = 2$, $UI_2U^T = (\det U)I_2 = I_2$ is constant, in accordance with the isomorphism $\mathrm{SU}(2) \cong \mathrm{Sp}(2)$.

[7] If the quark masses are non-degenerate, the $\Pi^a$ modes acquire masses.

$$A = \exp\left(i\frac{\eta'}{\sqrt{N_f}f_{\eta'}}\right). \tag{3.34}$$

Expanding to second order in terms of the $\eta'$ field in (3.29) yields

$$f_{\eta'}^2 m_{\eta'}^2 = 4c\Delta^2 m^2 . \tag{3.35}$$

Since $f_{\pi,\eta'} \sim \mu$ [4] we obtain from (3.33) and (3.35) the relation

$$m_{\Pi,\eta'} \sim \frac{m\Delta}{\mu} . \tag{3.36}$$

Next, we consider the simpler case $N_f = 2$ [34]. In this case, no exact symmetry is spontaneously broken by the diquark condensate (3.19), and only the breaking of $\mathrm{U}(1)_A$ is relevant. The first line of (3.29) requires no change while the second line is replaced by

$$-c'\Delta^2\Big\{(\det M)A^2 + \text{c.c.}\Big\} \quad \text{with} \quad c' = \frac{3}{2\pi^2} . \tag{3.37}$$

The above value of $c'$ corrects the value of $4/3\pi^2$ given in [34].[8]

Finally let me discuss the behavior of SU(2) gluons at high density. In two-color QCD, the diquark condensate (3.19) is a color singlet, so the SU(2) gluons do not acquire a mass through the Higgs mechanism, in contrast to the CFL phase where the gluons also acquire a rest mass of $\mathcal{O}(\Delta)$. The absence below $\Delta$ of particles charged under SU(2) indicates that the medium is transparent for SU(2) gluons, so that neither Debye screening nor Meissner effect occur [35]. For these reasons, the low-energy dynamics of gluons is simply described by the Lagrangian of SU(2) pure Yang-Mills theory. It has been found [36] that the confinement scale of in-medium SU(2) gluons is considerably diminished from the value $\sim\Lambda_{\mathrm{SU}(2)}$ at $\mu = 0$ to

$$\Lambda'_{\mathrm{SU}(2)} \equiv \Delta \exp\left(-\frac{2\sqrt{2}\pi}{11}\frac{\mu}{g\Delta}\right) \ll \Lambda_{\mathrm{SU}(2)} \tag{3.38}$$

due to the polarization effect of the diquark condensate. Since $\mu/g\Delta$ is large, it follows that gluons become almost gapless at asymptotically large $\mu$. However, as long as their coupling to NG bosons is small enough, one can neatly discard gluons from the chiral Lagrangian of NG bosons.

[8] We thank Thomas Schäfer for a communication on this point.

### 3.2.2 Finite-Volume Analysis

#### Microscopic Domain

In this subsection we introduce the $\varepsilon$-regime (or the microscopic domain) of dense two-color QCD. In the following, we will neglect the $H$ boson as well as gluons since they decouple from the dynamics of the other NG bosons and give no contribution to the light quark mass dependence of the partition function.

Let us study two-color QCD in a Euclidean four-dimensional torus of size $L \times L \times L \times \beta \equiv V_4$ with $\beta = 1/T \sim L$. The $\varepsilon$-regime is where the zero-momentum modes of NG bosons dominate the partition function [37]. In the present setting, this regime is defined by the inequalities

$$\frac{1}{\Delta} \ll L \ll \frac{1}{m_{\Pi,\tilde{\Pi},\eta'}}, \tag{3.39}$$

where $m_{\Pi,\tilde{\Pi},\eta'}$ is the mass of the NG bosons ($\Pi$, $\tilde{\Pi}$, $\eta'$) at large $\mu$. The first inequality, $1/\Delta \ll L$, means that the contribution of heavy non-NG bosons to the partition function can be neglected. The second inequality guarantees that the box size is much smaller than the Compton wave length of the NG bosons, so that the functional integral over NG boson fields reduces to the zero-mode integral over the coset space

$$\frac{\mathrm{SU}(N_f)_R \times \mathrm{SU}(N_f)_L \times \mathrm{U}(1)_A}{\mathrm{Sp}(N_f)_R \times \mathrm{Sp}(N_f)_L}.$$

#### Massless Spectral Sum Rules

In this subsection we determine the dependence of the finite-volume partition function on the light quark masses and use it to derive a set of exact spectral sum rules for the eigenvalues of the Dirac operator $\mathcal{D}(\mu)$. Let us first consider $N_f \geq 4$ (the case $N_f = 2$ will be discussed separately). Since the symmetry breaking pattern $\mathrm{SU}(n) \to \mathrm{Sp}(n)$ is the same as at $\mu = 0$ we can directly borrow calculational tools from [38]. The finite-volume partition function in the high density $\varepsilon$-regime (3.39) is given by

$$Z(M) = \int_{\mathrm{U}(1)_A} [dA] \int_{\mathrm{SU}(N_f)_L/\mathrm{Sp}(N_f)_L} [d\Sigma_L] \int_{\mathrm{SU}(N_f)_R/\mathrm{Sp}(N_f)_R} [d\Sigma_R] \; \exp\left[cV_4\Delta^2\left\{A^2 \operatorname{tr}[M\Sigma_R M^T \Sigma_L^\dagger] + \text{c.c.}\right\}\right] \tag{3.40}$$

$$= \int_{\mathrm{U}(1)_A} [dA] \int_{\mathrm{SU}(N_f)_L} [dU_L] \int_{\mathrm{SU}(N_f)_R} [dU_R] \; \exp\left[-cV_4\Delta^2\left\{A^2 \operatorname{tr}[MU_R I_{N_f} U_R^T M^T U_L^* I_{N_f} U_L^\dagger] + \text{c.c.}\right\}\right]. \tag{3.41}$$

In going from the first to the second line of (3.41), we have used $\Sigma = U I_{N_f} U^T$ as in (3.31), but for computational convenience we have chosen $U \in \mathrm{SU}(N_f)$ rather than $U \in \mathrm{SU}(N_f)/\mathrm{Sp}(N_f)$. The integral over $U$ is independent of the additional degrees of freedom we have introduced because $U I_{N_f} U^T = I_{N_f}$ for $U \in \mathrm{Sp}(N_f)$.

From the invariance property of the Haar measure, (3.41) can be expanded as

$$Z(M) = 1 + c_2 \operatorname{tr} M^\dagger M + c_4^{(1)} (\operatorname{tr} M^\dagger M)^2 + c_4^{(2)} \operatorname{tr}(M^\dagger M)^2 + \mathcal{O}\big(M^6\big) . \quad (3.42)$$

However, all terms of $\mathcal{O}(M^{4k-2})$ ($k = 1, 2, \ldots$) vanish because of the $\mathrm{U}(1)_A$ integral. In particular $c_2 = 0$, so the first nontrivial coefficients are $c_4^{(1)}$ and $c_4^{(2)}$, both proportional to $(cV_4\Delta^2)^2$. To figure out their proportionality constants, we use the formulas given in [[38], Eqs. (4.11)–(4.17)],

$$\int\limits_{\mathrm{U}(2N)} [dU] \exp\big(\operatorname{tr}[Y U I_{2N} U^T] + \text{c.c.}\big) = 1 + \frac{2}{2N-1} \operatorname{tr}[Y^\dagger Y] + \mathcal{O}(Y^4) , \quad (3.43)$$

which holds for any $2N \times 2N$ antisymmetric matrix $Y$. Comparison of (3.41) and (3.43) suggests the substitutions $U \to A U_R$, $N \to N_f/2$, and $Y \to -cV_4\Delta^2 M^T U_L^* I_{N_f} U_L^\dagger M$ ($Y$ is antisymmetric). We thereby obtain

$$\begin{aligned} Z(M) &= \int\limits_{\mathrm{SU}(N_f)_L} [dU_L] \left\{ 1 - (cV_4\Delta^2)^2 \frac{2}{N_f - 1} \operatorname{tr}[M^\dagger U_L I_{N_f} U_L^T M^* \cdot M^T U_L^* I_{N_f} U_L^\dagger M] \right\} + \mathcal{O}(M^8) \\ &= 1 + (cV_4\Delta^2)^2 \frac{2}{(N_f-1)^2} \left\{ (\operatorname{tr} M^\dagger M)^2 - \operatorname{tr}(M^\dagger M)^2 \right\} + \mathcal{O}(M^8) . \end{aligned} \quad (3.44)$$

We note in passing that the above combination $(\operatorname{tr}[M^\dagger M])^2 - \operatorname{tr}[(M^\dagger M)^2]$ also appeared in the previous section for the CFL phase, (3.11). The above expansion is to be matched later against the one derived directly from the fundamental QCD Lagrangian.

As a special case, let us consider the case of a flavor-symmetric mass term $M = m\mathbf{1}_{N_f}$ with real $m$. Now one of the two $\mathrm{SU}(N_f)$ integrals (which corresponds to the integral over the massless modes $\Pi^a$ in (3.33)) trivially drops out, leaving

$$Z(m) = \int\limits_{\mathrm{U}(N_f)} [dU] \exp\Big[ -cV_4\Delta^2 m^2 \Big( \operatorname{tr}[U I_{N_f} U^T I_{N_f}] + \text{c.c.} \Big) \Big] . \quad (3.45)$$

A small consistency check can be done using (3.43) in (3.45) to obtain (3.44) with $M = m\mathbf{1}_{N_f}$. This expression is of the same form as the one at $\mu = 0$ [[38], Eq. (4.11)], which can be rewritten as a pfaffian [[38], Eq. (5.13)]). By the same token (with the replacements $x \to 4cV_4\Delta^2 m^2$, $\nu \to 0$, and $2N_f \to N_f$, respectively, in [[38], Eq. (5.13)]) we readily obtain the simple expression[9]

---

[9] A pfaffian of an $n \times n$ antisymmetric matrix $X$ (for even $n$) is defined as

$$Z(m) = \frac{1}{(N_f - 1)!!} \,\mathrm{Pf}(A)\,, \tag{3.46}$$

where $A$ is an $N_f \times N_f$ antisymmetric matrix with entries

$$A_{pq} \equiv (q-p) I_{p+q}(4cV_4\Delta^2 m^2)\,, \quad p, q = -\frac{N_f - 1}{2}, \ldots, \frac{N_f - 3}{2}, \frac{N_f - 1}{2} \tag{3.47}$$

and $I_{p+q}$ denotes a modified Bessel function. Moreover an analytical closed formula for (3.41) for generic quark masses is available if we use the formula by Kieburg and Guhr [39] (Sect. 3.4.2).

Having determined the light quark mass dependence of the partition function, we can now derive spectral sum rules for two-color QCD at large $\mu$. Let us define

$$\langle \mathcal{O} \rangle \equiv \int [\mathcal{D}A]\, \mathcal{O} \Big( \prod_n{}' \lambda_n^2 \Big)^{N_f} \mathrm{e}^{-S_g} \Big/ \int [\mathcal{D}A] \Big( \prod_n{}' \lambda_n^2 \Big)^{N_f} \mathrm{e}^{-S_g}\,, \tag{3.48}$$

where $\prod_n'$ (and later $\sum_n'$) denotes the product (sum) over all eigenvalues $\lambda_n$ with $\mathrm{Re}\,\lambda_n > 0$, and $S_g \equiv \int d^4x\, F^a_{\mu\nu} F^a_{\mu\nu}/4$ is the gluonic action. Because of the pseudo-reality of SU(2), the measure in (3.48) is real and positive definite (for even $N_f$).

We now compare the finite-volume partition function (3.44) with the expression thereof in terms of Dirac eigenvalues $\{i\lambda_n\}$, as in the previous section, which leads to the spectral sum rules

$$\Big\langle \sum_n{}' \frac{1}{\lambda_n^2} \Big\rangle = \Big\langle \sum_{m<n}{}' \frac{1}{\lambda_m^2 \lambda_n^2} \Big\rangle = \Big\langle \sum_n{}' \frac{1}{\lambda_n^6} \Big\rangle = 0\,, \tag{3.49}$$

$$\Big\langle \sum_n{}' \frac{1}{\lambda_n^4} \Big\rangle = \frac{9}{4(N_f-1)^2\pi^4} \left( V_4 \Delta^2 \right)^2, \qquad \text{for } N_f \geq 4, \text{ even.} \tag{3.50}$$

We can derive infinitely many sum rules if we continue the expansion in terms of masses to arbitrarily higher orders.

The spectral sums appearing in (3.49) and (3.50) are real-valued. However they are real-valued even *before* the average over gauge fields, because of the the pseudo-reality of SU(2): if $\lambda$ is an eigenvalue of $\mathcal{D}(\mu)$, so is $\lambda^*$. Combined with the chiral symmetry $\{\gamma_5, \mathcal{D}(\mu)\} = 0$, this implies that the Dirac eigenvalues in dense two-color QCD occur in quadruplets $(\lambda_n, \lambda_n^*, -\lambda_n, -\lambda_n^*)$. When $\lambda_n$ coincides with $\pm\lambda_n^*$, the quadruplet degenerates into a doublet of *exactly real* or *purely imaginary* eigenvalues. Summarizing, all nonzero Dirac eigenvalues in dense two-color QCD are classified

---

$$\mathrm{Pf}(X) = \frac{1}{2^{n/2}(n/2)!} \sum_\sigma \mathrm{sgn}(\sigma) X_{\sigma(1)\sigma(2)} \cdots X_{\sigma(n-1)\sigma(n)}\,,$$

where the sum runs over all permutations $\sigma$ of $1, 2, \ldots, n$. Note that this definition differs by a factor of $1/2^{n/2}$ from [[38]Eq. (A.8)].

into three types: quadruplets of complex eigenvalues $(\lambda_n, \lambda_n^*, -\lambda_n, -\lambda_n^*)$, doublets of exactly real eigenvalues $(\lambda_n, -\lambda_n) \in \mathbb{R}$, and doublets of purely imaginary eigenvalues $(\lambda_n, -\lambda_n) \in i\mathbb{R}$. This complicated structure of the spectrum is a distinctive feature of the $\beta = 1$ symmetry class.

As a next step let us define the spectral density

$$\rho(\lambda) = \left\langle \sum_n \delta^2(\lambda - \lambda_n) \right\rangle, \tag{3.51}$$

Unlike in the CFL phase, this is a positive-definite function since the path-integral measure in the chiral limit of two-color QCD is non-negative definite. (The case of nonzero masses will be considered in Sect. 3.2.2).

With $\rho(\lambda)$, (3.50) may be expressed as

$$\frac{1}{(4cV_4\Delta^2)^2} \int_{\mathbb{C}_+} d^2\lambda \, \frac{\rho(\lambda)}{\lambda^4} = \int_{\mathbb{C}_+} d^2z \, \frac{\rho_s^{V_4}(z)}{z^4} = \frac{1}{4(N_f - 1)^2} \tag{3.52}$$

with $\mathbb{C}_+ \equiv \{w \in \mathbb{C} \mid \operatorname{Re} w > 0\}$ and

$$\rho_s^{V_4}(z) \equiv \frac{\pi^2}{3V_4\Delta^2} \, \rho\left(\frac{\pi z}{\sqrt{3V_4\Delta^2}}\right), \tag{3.53}$$

where the numerical factor $4c = 3/\pi^2$ has been included in the definition to simplify some analytical results in Sect. 3.2.2. Equation (3.52) suggests the existence of the *microscopic spectral density* defined by

$$\rho_s(z) \equiv \lim_{V_4 \to \infty} \rho_s^{V_4}(z). \tag{3.54}$$

Since the sum rule (3.52) is determined solely by the symmetry breaking pattern induced by the diquark condensate,[10] we expect that (3.54) is a *universal* function, i.e., fully determined by the global symmetries and independent of the details of UV interactions.

Our conjecture is corroborated by various facts known at $\mu = 0$ [40]:

- The Leutwyler-Smilga sum rules are satisfied to good accuracy by the Dirac eigenvalues computed in the instanton liquid model, a phenomenological model of QCD.
- The microscopic spectral correlation functions derived from ChRMT satisfy the Leutwyler-Smilga sum rules at $\mu = 0$ exactly.
- The microscopic spectral correlation functions from ChRMT have been verified in a number of lattice QCD simulations.

It is thus natural to expect that the way the thermodynamic limit of the spectral density near zero is approached is universal at large $\mu$, too. Further insights would

---

[10] Note that the mass term at $\mathcal{O}(M^2)$ is unique, in contrast to the CFL phase (cf. p. 66).

be obtained if we could construct the appropriate ChRMT, a topic we address in the next section.

It is interesting to ask how the vanishing sum rules (3.49) can be understood in terms of the symmetries of the distribution of eigenvalues. We turn to the spectral density $\rho(\lambda)$. It satisfies $\rho(\lambda) = \rho(\lambda^*) = \rho(-\lambda) = \rho(-\lambda^*)$, but this is not enough to prove the vanishing of sum rules. We now show that (3.49) follows automatically if $\rho(\lambda)$ also satisfies $\rho(\lambda) = \rho(i\lambda)$:

$$\int d^2\lambda \, \frac{\rho(\lambda)}{\lambda^{4k-2}} = \int d^2\lambda \, \frac{\rho(i\lambda)}{(i\lambda)^{4k-2}} = -\int d^2\lambda \, \frac{\rho(\lambda)}{\lambda^{4k-2}} = 0 \,. \tag{3.55}$$

In the next section we show that the exact microscopic spectral density obtained from ChRMT does indeed possess this property in the chiral limit (→ cf. (3.126)).

We finally consider $N_f = 2$. From (3.37) the partition function in the $\varepsilon$-regime is found to be

$$Z(M) = \int_{\mathrm{U}(1)_A} dA \, \exp\left\{ V_4 c' \Delta^2 \Big[ (\det M) A^2 + \text{c.c.} \Big] \right\} = I_0(2V_4 c' \Delta^2 |\det M|) \,. \tag{3.56}$$

Comparing (3.56) with the spectral representation of $Z(M)$, we again find the vanishing sum rules (3.49) as well as

$$\left\langle {\sum_n}' \frac{1}{\lambda_n^4} \right\rangle = \frac{9}{4\pi^4} \left( V_4 \Delta^2 \right)^2 , \tag{3.57}$$

which again hints at the existence of the universal microscopic spectral density (3.54). Note that (3.57) is actually the same as (3.50) with $N_f = 2$. In other words, the analysis for $N_f \geq 4$ pertains to $N_f = 2$ if we set $\Sigma_L = \Sigma_R = I_2$ (corresponding to the trivial coset space SU(2)/Sp(2)) and $U_L = U_R = \mathbf{1}_2$ in (3.41) and (3.44).

### Massive Spectral Sum Rules

It is known that the massless spectral sum rules at $\mu = 0$ due to Leutwyler and Smilga can be generalized to the massive case by an appropriate rescaling of masses [13], and that the double-microscopic spectral correlations (called so because masses and eigenvalues are rescaled simultaneously as the volume is taken to infinity) derived from ChRMT reproduce these sum rules [41–43]. Here we show that the new spectral sum rules derived at large $\mu$ in the chiral limit can be generalized to the massive case in a similar fashion. Such an extension may prove useful in future lattice simulations, because massless fermions are computationally quite expensive.

For simplicity we consider $N_f = 2$ with masses $m_1$ and $m_2$. According to the last section, the partition function in the $\varepsilon$-regime reads

$$\begin{aligned} Z &= I_0(\alpha m_1 m_2) \quad \text{with} \quad \alpha = \frac{3V_4\Delta^2}{\pi^2} \\ &= \frac{1}{\mathcal{N}} \int [\mathcal{D}A] \prod_n{}' [(\lambda_n^2 + m_1^2)(\lambda_n^2 + m_2^2)] \mathrm{e}^{-S_g}, \end{aligned} \tag{3.58}$$

where $\mathcal{N}$ is a mass-independent normalization factor. In the remainder of this section we will denote the expectation value with respect to this measure by $\langle\langle \cdots \rangle\rangle$.

Differentiation of $Z$ by the masses yields massive spectral sum rules, e.g.,[11]

$$\frac{\partial}{\partial(m_1^2)} \log Z = \left\langle\left\langle \sum_n{}' \frac{1}{\lambda_n^2 + m_1^2} \right\rangle\right\rangle = \frac{(\alpha m_2)^2}{4} \frac{I_0(\alpha m_1 m_2) - I_2(\alpha m_1 m_2)}{I_0(\alpha m_1 m_2)}, \tag{3.59}$$

$$\frac{1}{Z} \frac{\partial^2 Z}{\partial(m_1^2)\partial(m_2^2)} = \left\langle\left\langle \sum_n{}' \frac{1}{\lambda_n^2 + m_1^2} \sum_k{}' \frac{1}{\lambda_k^2 + m_2^2} \right\rangle\right\rangle = \frac{\alpha^2}{4}, \tag{3.60}$$

$$\begin{aligned} \frac{1}{Z} \frac{\partial^2 Z}{\partial(m_1^2)^2} &= \left\langle\left\langle \sum_{n \neq k}{}' \frac{1}{(\lambda_n^2 + m_1^2)(\lambda_k^2 + m_1^2)} \right\rangle\right\rangle \\ &= \frac{(\alpha m_2)^4}{96} \frac{3I_0(\alpha m_1 m_2) - 4I_2(\alpha m_1 m_2) + I_4(\alpha m_1 m_2)}{I_0(\alpha m_1 m_2)}. \end{aligned} \tag{3.61}$$

For degenerate masses ($m_1 = m_2 \equiv m$), $Z = I_0(\alpha m^2)$ so that

$$\frac{1}{Z} \frac{\partial^2 Z}{\partial(m^2)^2} = \left\langle\left\langle -2\sum_n{}' \frac{1}{(\lambda_n^2 + m^2)^2} + 4\Big(\sum_n{}' \frac{1}{\lambda_n^2 + m^2}\Big)^2 \right\rangle\right\rangle = \frac{\alpha^2}{2} \frac{I_0(\alpha m^2) + I_2(\alpha m^2)}{I_0(\alpha m^2)}. \tag{3.62}$$

In the massless limit (3.59)–(3.62) are consistent with the massless sum rules (3.49) and (3.57), as they should be. In terms of rescaled dimensionless variables,

$$z_n \equiv \lambda_n \frac{\sqrt{3V_4\Delta^2}}{\pi}, \qquad \mathbf{m}_i \equiv m_i \frac{\sqrt{3V_4\Delta^2}}{\pi}, \tag{3.63}$$

(3.59)–(3.62) become

$$\left\langle\left\langle \sum_n{}' \frac{1}{z_n^2 + \mathbf{m}_1^2} \right\rangle\right\rangle = \frac{\mathbf{m}_2^2}{4} \frac{I_0(\mathbf{m}_1\mathbf{m}_2) - I_2(\mathbf{m}_1\mathbf{m}_2)}{I_0(\mathbf{m}_1\mathbf{m}_2)}, \tag{3.64}$$

$$\left\langle\left\langle \sum_n{}' \frac{1}{z_n^2 + \mathbf{m}_1^2} \sum_k{}' \frac{1}{z_k^2 + \mathbf{m}_2^2} \right\rangle\right\rangle = \frac{1}{4}, \tag{3.65}$$

$$\left\langle\left\langle \sum_{n \neq k}{}' \frac{1}{(z_n^2 + \mathbf{m}_1^2)(z_k^2 + \mathbf{m}_1^2)} \right\rangle\right\rangle = \frac{\mathbf{m}_2^4}{96} \frac{3I_0(\mathbf{m}_1\mathbf{m}_2) - 4I_2(\mathbf{m}_1\mathbf{m}_2) + I_4(\mathbf{m}_1\mathbf{m}_2)}{I_0(\mathbf{m}_1\mathbf{m}_2)}, \tag{3.66}$$

---

[11] We repeatedly used the identities $I_n'(x) = [I_{n-1}(x) + I_{n+1}(x)]/2$ and $I_n(x)/x = [I_{n-1}(x) - I_{n+1}(x)]/2n$.

$$\left\langle\!\!\left\langle -\sum_n{}' \frac{1}{(z_n^2+\mathbf{m}^2)^2} + 2\Big(\sum_n{}' \frac{1}{z_n^2+\mathbf{m}^2}\Big)^2 \right\rangle\!\!\right\rangle = \frac{I_0(\mathbf{m}^2)+I_2(\mathbf{m}^2)}{4I_0(\mathbf{m}^2)} . \tag{3.67}$$

It is surprising that the right-hand side of (3.65) is independent of the masses.

We now define the double-microscopic spectral density for even $N_f$ by

$$\rho_s^{(N_f)}(z;\{\mathbf{m}_i\}) \equiv \lim_{V_4\to\infty} \frac{\pi^2}{3V_4\Delta^2}\rho\left(\frac{\pi z}{\sqrt{3V_4\Delta^2}}\right)\Bigg|_{\mathbf{m}_i=m_i\frac{\sqrt{3V_4\Delta^2}}{\pi}\ \text{fixed}} . \tag{3.68}$$

From (3.65) and (3.67) we can derive a formula fulfilled by the massive spectral density for $N_f=2$:

$$\int_{\mathbb{C}_+} d^2z\, \frac{\rho_s^{(2)}(z;\mathbf{m},\mathbf{m})}{(z^2+\mathbf{m}^2)^2} = \frac{I_0(\mathbf{m}^2)-I_2(\mathbf{m}^2)}{4I_0(\mathbf{m}^2)} . \tag{3.69}$$

In Sect. 3.4 we will derive $\rho_s^{(N_f)}(z;\{\mathbf{m}_i\})$ analytically based on the exact mapping to ChRMT.

For larger $N_f$ the explicit expressions become increasingly involved. For $N_f=4$ with equal masses, we have

$$\left\langle\!\!\left\langle \sum_n{}' \frac{1}{z_n^2+\mathbf{m}^2} \right\rangle\!\!\right\rangle = \frac{2I_0(\mathbf{m}^2)I_1(\mathbf{m}^2) - 3I_1(\mathbf{m}^2)I_2(\mathbf{m}^2) + I_2(\mathbf{m}^2)I_3(\mathbf{m}^2)}{4Z} , \tag{3.70}$$

where

$$Z = 3I_0(\mathbf{m}^2)^2 - 4I_1(\mathbf{m}^2)^2 + 3I_2(\mathbf{m}^2)^2 . \tag{3.71}$$

Analogously,

$$\left\langle\!\!\left\langle -\sum_n{}' \frac{1}{(z_n^2+\mathbf{m}^2)^2} + 4\Big(\sum_n{}' \frac{1}{z_n^2+\mathbf{m}^2}\Big)^2 \right\rangle\!\!\right\rangle = \frac{2I_0^2+I_1^2-2I_2^2+I_3^2-I_0I_2-2I_1I_3+I_2I_4}{8Z} . \tag{3.72}$$

One can check that the massless sum rules (3.49) and (3.50) are reproduced correctly in the chiral limit. Many more sum rules can be derived from the partition function for different fermion masses although we do not work out them explicitly. We only mention that the sum rules corresponding to different masses satisfy non-trivial consistency conditions as some of the masses are sent to infinity, owing to decoupling [41].

## 3.3 ChRMT for Dense Two-Color QCD: Construction

In the last section it was shown that

1. Dense two-color QCD at high density (with even $N_f$) exhibits a symmetry breaking pattern

$$\mathrm{SU}(N_f)_L \times \mathrm{SU}(N_f)_R \times \mathrm{U}(1)_B \times \mathrm{U}(1)_A \to \mathrm{Sp}(N_f)_L \times \mathrm{Sp}(N_f)_R, \tag{3.73}$$

   which differs from the one at low density, $\mathrm{SU}(2N_f) \to \mathrm{Sp}(2N_f)$, owing to the large *explicit* symmetry breaking due to $\mu$. The diquark condensate itself has the same quantum numbers in both limits.
2. The $\varepsilon$-regime at high density can be defined, based on the static part of the chiral Lagrangian descended from (3.73). The relevant scale involved is the BCS gap $\Delta$, playing the role of $\langle \bar{q}q \rangle$ at low density.
3. The Leutwyler-Smilga-type spectral sum rules for Dirac eigenvalues indicate that the Dirac eigenvalue distribution on the scale $\mathcal{O}(1/\sqrt{V_4\Delta^2})$ will be universal and allow for a description by chiral random matrix theory (ChRMT).

In this section we will argue, following [44], that

$$Z_{\mathrm{RMT}}(\{m_f\}) \equiv \int_{\mathbb{R}} \prod_{i,j} dA_{ij}\, dB_{ij} \prod_{f=1}^{N_f} \det \begin{pmatrix} m_f \mathbf{1}_N & A \\ B^T & m_f \mathbf{1}_{N+\nu} \end{pmatrix} \mathrm{e}^{-N\,\mathrm{tr}[AA^T + BB^T]}, \tag{3.74}$$

is the sought-after ChRMT that corresponds to dense two-color QCD with even $N_f$.[12] Here $A$ and $B$ are real $N \times (N+\nu)$ matrices and $\{m_f\}$ are dimensionless mass parameters. The pivotal properties of $Z_{\mathrm{RMT}}$ are as follows:

(i) The Dirac matrix $\mathcal{D} \equiv \begin{pmatrix} \mathbf{0} & A \\ B^T & \mathbf{0} \end{pmatrix}$ has $|\nu|$ exact zero modes, respects chiral symmetry $\{\gamma_5, \mathcal{D}\} = 0$, and is *maximally non-Hermitian* in the sense that off-diagonal blocks fluctuate independently (cf. p. 47).
(ii) If $\lambda$ is an eigenvalue of $\mathcal{D}$, so are $-\lambda$, $\lambda^*$ and $-\lambda^*$.
(iii) $Z_{\mathrm{RMT}}(\{{}^\forall m_f = 0\})$ is invariant under $\mathrm{U}(N_f) \times \mathrm{U}(N_f)$, which is broken spontaneously down to $\mathrm{Sp}(N_f) \times \mathrm{Sp}(N_f)$ in the large-$N$ limit.

The property (i) suggests that $\nu$ corresponds to the topological charge in QCD. Since the topological susceptibility is strongly suppressed at high density [10, 46] one can set $\nu = 0$.

The property (ii) follows from the off-diagonal block structure and the reality of $\mathcal{D}$. It is implied that all nonzero eigenvalues come either in quadruplet $(\lambda, -\lambda, \lambda^*, -\lambda^*)$

[12] The model (3.74) is not new. It corresponds to class $2P$ in Magnea's mathematical classification of non-Hermitian random matrix ensembles, see ([[45] Table 2]). What is new is the realization that (3.74) describes dense two-color QCD.

with $\lambda \in \mathbb{C} \setminus \mathbb{R}$ or in doublet $(\lambda, -\lambda)$ with $\lambda \in \mathbb{R}$. This property is the hallmark of the Dirac spectrum in two-color QCD, as explained in the last section (cf. page 41).

The property (iii) may be seen as follows. Let us rewrite $Z_{\text{RMT}}(\{{}^{\forall}m_f = 0\})$ as follows:

$$Z_{\text{RMT}}(\{0\}) = \int [dA] \det{}^{N_f/2} \begin{pmatrix} \mathbf{0} & A \\ -A^T & \mathbf{0} \end{pmatrix} \mathrm{e}^{-N \operatorname{tr}[AA^T]} \\ \times \int [dB] \det{}^{N_f/2} \begin{pmatrix} \mathbf{0} & B \\ -B^T & \mathbf{0} \end{pmatrix} \mathrm{e}^{-N \operatorname{tr}[BB^T]}. \qquad (3.75)$$

Each of the two factors on the RHS is equal to the (massless) partition function of **the Chiral Gaussian Orthogonal Ensemble (ChGOE)** at $\mu = 0$ for $N_f/2$ flavors (see (2.80) and (2.81)), for which the flavor symmetry breaking pattern in the large-$N$ limit is known to be $\mathrm{U}(N_f) \to \mathrm{Sp}(N_f)$ [47, 48].[13] Thus we can immediately conclude that $Z_{\text{RMT}}$ exhibits the symmetry breaking $\mathrm{U}(N_f) \times \mathrm{U}(N_f) \to \mathrm{Sp}(N_f) \times \mathrm{Sp}(N_f)$, which agrees with (3.73).

To figure out the precise correspondence between the RMT parameters and the physical masses, we study the large-$N$ limit of $Z_{\text{RMT}}$ explicitly. Following the procedure reviewed in Chap. 2 Sect. 2.3.3, we are led to

$$Z_{\text{RMT}}(M) = \int_{\mathbb{C}} \prod_{i<j} dK_{ij}\, dL_{ij}\, \mathrm{e}^{-N \operatorname{tr}[KK^\dagger + LL^\dagger]} \operatorname{Pf}^N \begin{pmatrix} L^\dagger & -M^* \\ M^\dagger & K \end{pmatrix} \operatorname{Pf}^N \begin{pmatrix} K^\dagger & -M^T \\ M & L \end{pmatrix}, \qquad (3.76)$$

where $K$ and $L$ are $N_f \times N_f$ complex *antisymmetric* matrices and $M$ denotes the mass matrix, as usual. Here $\nu = 0$ is assumed. This is an exact transformation.

Since $K$ and $L$ are antisymmetric they can be brought to the standard form $K = U \Lambda U^T$ with $U \in \mathrm{U}(N_f)/[\mathrm{Sp}(2)]^{N_f/2}$ and $\Lambda$ a real antisymmetric matrix with $\Lambda_{k,k+1} = -\Lambda_{k+1,k} \geq 0$ and all other matrix elements zero (likewise for $L$). For infinitesimal $\|M\|$, the integration over $\Lambda$ can be estimated with a saddle-point approximation whereas the integration over $U$ is soft and has to be done exactly. A short calculation shows that the saddle point is located at $\Lambda = I_{N_f}/\sqrt{2}$ with $I_{N_f} \equiv \begin{pmatrix} \mathbf{0} & -\mathbf{1}_{N_f/2} \\ \mathbf{1}_{N_f/2} & \mathbf{0} \end{pmatrix}$ (defined in Chap. 2 Eq. (2.17)). Since $U$ enters $Z_{\text{RMT}}$ only in the form $U I_{N_f} U^T$, the integration manifold can be enlarged to $\mathrm{U}(N_f)$. Thus

[13] To avoid confusion: On page 41 the pattern is denoted as $\mathrm{SU}(2N_f) \to \mathrm{Sp}(2N_f)$ because $\mathrm{U}(1)_A$ is not regarded as an exact symmetry there.

$$
\begin{aligned}
Z_{\mathrm{RMT}}(M) &\sim \iint\limits_{\mathrm{U}(N_f)} [dU]\,[dV]\,\mathrm{Pf}^N \begin{pmatrix} \frac{1}{\sqrt{2}}(V I_{N_f} V^T)^\dagger & -M^* \\ M^\dagger & \frac{1}{\sqrt{2}} U I_{N_f} U^T \end{pmatrix} \\
&\quad \times \mathrm{Pf}^N \begin{pmatrix} \frac{1}{\sqrt{2}}(U I_{N_f} U^T)^\dagger & -M^T \\ M & \frac{1}{\sqrt{2}} V I_{N_f} V^T \end{pmatrix} \\
&\sim \iint\limits_{\mathrm{U}(N_f)} [dU]\,[dV]\, \exp\big( -2N \,\mathrm{Re}\,\mathrm{tr}[M U I_{N_f} U^T M^T V^* I_{N_f} V^\dagger] \big)\,, \qquad (3.77)
\end{aligned}
$$

where we expanded the exponent to the first nontrivial order in $M$. *This is identical to the finite-volume partition function at large* $\mu$, (3.41), *if we identify*

$$
N m_f^2 \Big|_{\mathrm{RMT}} \Longleftrightarrow \frac{3}{4\pi^2} V_4 \Delta^2 m_f^2 \Big|_{\mathrm{QCD}}\,. \qquad (3.78)
$$

The fact that $Z_{\mathrm{QCD}}$ in the $\varepsilon$-regime and $Z_{\mathrm{RMT}}$ have identical quark mass dependence implies that the Leutwyler-Smilga-type spectral sum rules descending from the Taylor expansion thereof, are *identical for the Dirac eigenvalues of QCD and for the eigenvalues of the random matrix $\mathcal{D}$ in RMT.* This strongly suggests that (3.74) is the right ChRMT for dense two-color QCD, paving the way toward a quantitative understanding of the Dirac spectrum at large $\mu$. A full proof of the equivalence, which would also establish the equality of all spectral correlation functions, requires the usage of the partially quenched effective theory, which is deferred to future work.

It should be stressed that the model (3.74) differs intrinsically from earlier phenomenological applications of RMT to dense QCD [49–54]. In these approaches, the corresponding low-energy effective theory was formulated on the basis of the symmetry breaking pattern at $\mu = 0$ [55], and $\mu$ was considered as a small perturbation. On the other hand, the low-energy effective theory corresponding to (3.74) is formulated in terms of NG bosons parameterizing the coset space $\mathrm{SU}(N_f) \times \mathrm{SU}(N_f)/(\mathrm{Sp}(N_f) \times \mathrm{Sp}(N_f))$, respecting the pattern of spontaneous symmetry breaking induced by the BCS-type diquark condensation at high density, with a vanishing chiral condensate. Consequently *our model includes no external parameter that stands for* $\mu$: (3.74) depends on $\mu$ only implicitly through $\Delta$ (cf. $\Delta \sim \mu \mathrm{e}^{-1/g}$ [56]).

Our model (3.79) corresponds to two-color QCD at high density, and cannot be used at small $\mu$. For this purpose we must use another random matrix model

$$
Z^{\mathrm{RMT}}_{N_f,\nu}(\hat{\mu}, \{m_f\}) = \int\limits_{\mathbb{R}} \prod_{i,j} dC_{ij}\, dD_{ij} \prod_{f=1}^{N_f} \det \begin{pmatrix} m_f \mathbf{1}_N & C + \hat{\mu} D \\ -C^T + \hat{\mu} D^T & m_f \mathbf{1}_{N+\nu} \end{pmatrix} \mathrm{e}^{-2N\,\mathrm{tr}[CC^T + DD^T]}\,, \qquad (3.79)
$$

where $C$ and $D$ are real $N \times (N+\nu)$ matrices and $\hat{\mu}$ is a parameter corresponding to the physical chemical potential $\mu$. This model was introduced in [57] as an extension of Osborn's two-matrix model of three-color QCD (2.92) to two colors. It is called

**the chiral real Ginibre ensemble** because it can also be seen as a chiral extension of the real Ginibre ensemble. It should be emphasized that it is not a phenomenological model of two-color QCD but is expected to be an exact theory of universal spectral correlations in the low end of the Dirac spectrum. In [58] all spectral correlation functions of this model for $N_f = 0$ have been worked out in full details. However, a precise understanding of the link between the model (3.79) and two-color QCD at low $\mu$ has not been given so far. Below we will show that (3.79) indeed reduces to two-color QCD in the $\varepsilon$-regime at small $\mu$ in the limit of weak non-Hermiticity ($N \to \infty$ with $N\hat{\mu}^2$ and $Nm_f$ fixed).[14] This correspondence does not come as a surprise since we already know the link between three-color QCD at low $\mu$ and the weak limit of Osborn's model, as reviewed in Chap. 2 Sect. 2.3.4.

In the following discussion $N_f$ is not restricted to be even. Assuming $0 \le \hat{\mu} \le 1$, and following the procedure reviewed in p. 42, we find

$$Z^{\mathrm{RMT}}_{N_f,\nu}(\hat{\mu},\, M) = \int_{\mathbb{C}} \prod_{i,j} dK_{ij}\, dL_{ij}\, dP_{ij}\, \mathrm{e}^{-8N\,\mathrm{tr}[KK^\dagger + LL^\dagger + 2PP^\dagger]}$$
$$\times \mathrm{Pf}^N \begin{pmatrix} \sqrt{1+\hat{\mu}^2}L^\dagger & -\sqrt{1-\hat{\mu}^2}P^T - \frac{1}{2}M^* \\ \sqrt{1-\hat{\mu}^2}P + \frac{1}{2}M^\dagger & \sqrt{1+\hat{\mu}^2}K \end{pmatrix}$$
$$\times \mathrm{Pf}^{N+\nu} \begin{pmatrix} \sqrt{1+\hat{\mu}^2}K^\dagger & -\sqrt{1-\hat{\mu}^2}P^* - \frac{1}{2}M^T \\ \sqrt{1-\hat{\mu}^2}P^\dagger + \frac{1}{2}M & \sqrt{1+\hat{\mu}^2}L \end{pmatrix}, \tag{3.80}$$

where $K$, $L$ and $P$ are complex $N_f \times N_f$ matrices, with $K$ and $L$ being antisymmetric. This is an exact transformation. We define

$$\mathfrak{D} \equiv \begin{pmatrix} K^\dagger & -P^* \\ P^\dagger & L \end{pmatrix} \quad \text{and} \quad \mathfrak{M} \equiv \begin{pmatrix} \mathbf{0} & -M^T \\ M & \mathbf{0} \end{pmatrix}. \tag{3.81}$$

For *both* $\hat{\mu}^2$ and $M$ treated as infinitesimal parameters, the saddle point is located at $\hat{\mu} = 0$ and $M = \mathbf{0}$, i.e., $\mathfrak{D} = UI_{2N_f}U^T/4$ with $U \in \mathrm{U}(2N_f)$. Taylor expansion in $\hat{\mu}^2$ yields

$$\mathrm{Pf} \begin{pmatrix} \sqrt{1+\hat{\mu}^2}K^\dagger & -\sqrt{1-\hat{\mu}^2}P^* - \frac{1}{2}M^T \\ \sqrt{1-\hat{\mu}^2}P^\dagger + \frac{1}{2}M & \sqrt{1+\hat{\mu}^2}L \end{pmatrix}$$
$$\sim \det U\, \exp\left(4\hat{\mu}^2\,\mathrm{tr}[\mathfrak{D}\mathfrak{B}^T\mathfrak{D}^\dagger\mathfrak{B}] + 4\,\mathrm{tr}[\mathfrak{D}^\dagger\mathfrak{M}]\right), \tag{3.82}$$

where we used $\mathfrak{D}^\dagger\mathfrak{D} = \mathbf{1}_{2N_f}/16$ and introduced the $2N_f \times 2N_f$ baryon charge matrix $\mathfrak{B} \equiv \begin{pmatrix} \mathbf{1}_{N_f} & \mathbf{0} \\ \mathbf{0} & -\mathbf{1}_{N_f} \end{pmatrix}$. Isolating the U(1) part by $U \to U\mathrm{e}^{i\theta/2N_f}$ we find

[14] See p. 39.

$$Z^{\mathrm{RMT}}_{N_f,\nu}(\hat{\mu}, M) \sim \int [d\theta]\, \mathrm{e}^{i\nu\theta} \int_{\mathrm{SU}(2N_f)} [dU]\, \exp\Big(\frac{1}{2}N\hat{\mu}^2\, \mathrm{tr}[U I_{2N_f} U^T \mathfrak{B}^T (U I_{2N_f} U^T)^\dagger \mathfrak{B}] + 2N\, \mathrm{Re}\, \mathrm{tr}[\mathrm{e}^{i\theta/N_f} U I_{2N_f} U^T \mathfrak{M}^\dagger]\Big)\,. \tag{3.83}$$

The exponent in this partition function *exactly reproduces the static part of the chiral Lagrangian at small* $\mu$ (2.32) obtained from symmetry principles, upon the identification

$$\frac{1}{2}N\hat{\mu}^2\Big|_{\mathrm{RMT}} \Longleftrightarrow V_4 F_\pi^2 \mu^2\Big|_{\mathrm{QCD}} \quad \text{and} \tag{3.84}$$

$$2Nm\Big|_{\mathrm{RMT}} \Longleftrightarrow V_4 F_\pi^2 m_\pi^2 = \frac{V_4 |\langle\bar{\psi}\psi\rangle| m_f}{2N_f}\Big|_{\mathrm{QCD}}\,, \tag{3.85}$$

where in the last equality the Gell-Mann–Oakes–Renner relation was used. A similar analysis was given in [[54], Eq. 4.15] based on the one-matrix formulation.

We have described two different matrix models that correspond to the high density and the low density of two-color QCD respectively. Unexpectedly, (3.79) reduces to (3.74) when $\nu = 0$ and $\hat{\mu} = 1$.[15] It is intriguing that a *single* ChRMT, (3.79), can describe two extreme cases, $\mu \gg \Lambda_{\mathrm{QCD}}$ and $\mu \sim m_\pi \ll \Lambda_{\mathrm{QCD}}$, that have totally distinct patterns of spontaneous symmetry breaking, by two different choices of the parameter ($\hat{\mu} \sim \mathcal{O}(1/\sqrt{N})$ and $\hat{\mu} = 1$, respectively) and two different mappings of the random-matrix quark masses to the physical quark masses (rescaling by $|\langle\bar{\psi}\psi\rangle|$ and $\Delta$, respectively).

A brief comment is in order concerning odd $N_f$. With $\hat{\mu} = 1$, assuming $M = m\mathbf{1}_{N_f}$ for simplicity, and allowing for $\nu \neq 0$ again, we find for the pfaffians in the integrand of (3.76)

$$\mathrm{Pf}^N \begin{pmatrix} L^\dagger & -M^* \\ M^\dagger & K \end{pmatrix} \mathrm{Pf}^{N+\nu} \begin{pmatrix} K^\dagger & -M^T \\ M & L \end{pmatrix} \propto \det{}^{N/2}(m^{*2}\mathbf{1}_{N_f} + L^\dagger K) \cdot \det{}^{(N+\nu)/2}(m^2 \mathbf{1}_{N_f} + LK^\dagger)\,. \tag{3.86}$$

Since $K$ (or $L$) is an antisymmetric matrix of odd size, at least one of its eigenvalues must be zero, and evidently the same holds for $L^\dagger K$ and $LK^\dagger$. Thus $Z^{\mathrm{RMT}}_\nu(1, M) \propto |m|^{2N} m^\nu$, and hence the model with odd $N_f$ seems to have an ill-defined chiral limit, with a divergent chiral condensate:

$$\frac{1}{N}\frac{\partial}{\partial m} \log Z^{\mathrm{RMT}}_{N_f,\nu}(1, M) \propto \frac{1}{m} \to \infty \qquad \text{as} \;\; m \to 0\,. \tag{3.87}$$

[15] The large-$N$ limit with $\mathcal{O}(1)$ non-Hermitian parameter is called *the limit of strong non-Hermiticity* in the conventional classification (p. 47).

Therefore the limit of strong non-Hermiticity cannot be taken for odd $N_f$. Further analysis on the high density limit of odd $N_f$ two-color QCD is left for future work.

In addition to the mass term, we could also add a diquark source term $\propto j\psi\psi$ to the model (3.74). It can be shown that the partition function at $N \gg 1$ then contains a term linear in $j$, resulting in a non-vanishing diquark condensate in this model. The addition of the diquark source term enables us to probe the spectral correlations of the singular values of the Dirac operator [59].

By now the physical value of (3.79) is made clear: the large-$N$ limit of strong (weak) non-Hermiticity gives us information on the low-lying Dirac eigenvalues at high (low) density. Next important task is to solve the model in both limits and derive the universal spectral density of Dirac eigenvalues explicitly. It is carried out in the next section.

## 3.4 ChRMT for Dense Two-Color QCD: Solution

Many years ago Ginibre introduced three ensembles of non-Hermitian random matrices, classified by $\beta = 1, 2$ and 4. The case $\beta = 1$, the Gaussian ensemble of real asymmetric matrices, turned out to be the most difficult one because the joint eigenvalue distribution function is not absolutely continuous; for finite $N$, there is always nonzero probability of finding *real* eigenvalues in addition to complex eigenvalues occurring in complex conjugate pairs. The complete solution of all real and complex eigenvalue correlations was obtained only recently by several groups [60–65]. Among others, Sommers and Wieczorek [60, 65] developed an elegant approach which we tentatively call *the variational approach*, but it was not 'closed' in itself, because they used Edelman's preceding result to calculate the kernel from which all eigenvalue correlations follow. Later Akemann, Phillips and Sommers [57] pointed out that the calculation of the kernel is reduced to that of an expectation value of a product of two characteristic polynomials. With this breakthrough the variational approach has become a closed, self-contained framework.

In [57, 58] the chiral extension of real Ginibre ensemble was given, with its potential application to two-color QCD at nonzero chemical potential. In the chiral real Ginibre ensemble, nonzero spectral density emerges not only on the real axis but also on the imaginary axis, in addition to the pairwise complex eigenvalues. In [57] the expectation value of characteristic polynomials was calculated for the chiral extension. In [58] the joint probability distribution function of eigenvalues was determined, and by combining these results within the variational approach, Akemann et al. obtained all eigenvalue correlation functions for the quenched case, at both limits of weak and strong non-Hermiticity. (To be precise, they encountered a subtle problem of convergence for the spectral density of real and pure imaginary eigenvalues, which was left unsolved). It is notable that they first determined the joint probability distribution function for the *square* of eigenvalues, and then converted it to that of the eigenvalues. In this way they managed to handle the chiral real Ginibre ensemble analogously to the real Ginibre ensemble.

The goal of this section is to achieve the generalization of the results in [58, 66] to $N_f > 0$ in the large-$N$ limit. We will obtain the microscopic spectral density in the large-$N$ limit of both the weak and strong non-Hermiticity, which corresponds to the low and high density limit of two-color QCD as shown in the last section. Our treatment essentially hinges on a mathematical formula by Kieburg and Guhr [39, 67] which enables us to calculate the large-$N$ partition function for an arbitrary number of flavors.

### 3.4.1 Spectral Density at Finite N

For a technical reason we first rewrite the partition function (3.79) in terms of new variables

$$\boxed{\tau_f = 2\sqrt{N} m_f}\,, A = 2\sqrt{N}(C + \hat{\mu} D),\ \ B = 2\sqrt{N}(C - \hat{\mu} D) \text{ and } \eta_\pm = \frac{1 \pm \hat{\mu}^2}{4\hat{\mu}^2}\,, \tag{3.88}$$

with which we obtain

$$\begin{aligned} & Z^{\mathrm{RMT}}_{N_f,\nu}(\hat{\mu}, \{\tau_f\} \\ &= \iint [dA]\,[dB] \prod_{f=1}^{N_f} \det\begin{pmatrix} \tau_f \mathbf{1}_N & A \\ -B^T & \tau_f \mathbf{1}_{N+\nu} \end{pmatrix} \mathrm{e}^{-\frac{1}{2}\eta_+ \operatorname{tr}(AA^T + BB^T) + \frac{1}{2}\eta^- \operatorname{tr}(AB^T + BA^T)} \end{aligned} \tag{3.89}$$

$$= \Big(\prod_{f=1}^{N_f} \tau_f^\nu\Big) \iint [dA]\,[dB] \prod_{f=1}^{N_f} \det(\tau_f^2 \mathbf{1}_N + AB^T) \mathrm{e}^{-\frac{1}{2}\eta_+ \operatorname{tr}(AA^T + BB^T) + \frac{1}{2}\eta^- \operatorname{tr}(AB^T + BA^T)}. \tag{3.90}$$

The integral for $A$ and $B$ is over $N \times (N + \nu)$ real matrices. The prefactor $\prod_{f=1}^{N_f} \tau_f^\nu$ may be omitted, because it drops out in all expectation values.

Let $z_k = x_k + i y_k \in \mathbb{C}\,(k = 1, \ldots, N)$ denote the $N$ eigenvalues of $AB^T$. They are related to the $2N$ nonzero eigenvalues $\{\Lambda_1, -\Lambda_1, \ldots, \Lambda_N, -\Lambda_N\}$ of the Dirac matrix

$$\mathcal{D}(\hat{\mu}) \equiv \begin{pmatrix} \mathbf{0} & A \\ -B^T & \mathbf{0} \end{pmatrix} = 2\sqrt{N} \begin{pmatrix} \mathbf{0} & C + \hat{\mu} D \\ -C^T + \hat{\mu} D^T & \mathbf{0} \end{pmatrix}, \tag{3.91}$$

as

$$z_k = -\Lambda_k^2\,. \tag{3.92}$$

We will first derive the spectral density for $\{z_k\}$ and then convert it to the spectral density of $\{\Lambda_k\}$. Since $AB^T$ is a real matrix, all eigenvalues must be either complex-conjugate pairs or unpaired exactly real eigenvalues. This observation is important in all developments in the following. Denoting the joint probability distribution function of $\{z_k\}$ by $d\Omega_{N,\nu}$ we have, by definition,

$$Z^{\mathrm{RMT}}_{N_f,\nu}(\hat{\mu},\{\tau_f\}) = \int d\Omega_{N,\nu} \prod_{f=1}^{N_f}\prod_{k=1}^{N}(\tau_f^2 + z_k)\,. \tag{3.93}$$

Note that $d\Omega_{N,\nu}$ does not depend on $N_f$ and the masses. In [58] it was determined to be

$$\begin{aligned}\int d\Omega_{N,\nu} = c_N \prod_{k=1}^{N}\int_{\mathbb{C}} d^2 z_k\, w_\nu(\hat{\mu}, z_k)\prod_{i<j}(z_i - z_j)\\ \times \sum_{n=0}^{[N/2]}\left(\left\{\prod_{l=1}^{n}(-2i)\delta(x_{2l-1}-x_{2l})\delta(y_{2l-1}+y_{2l})\Theta(y_{2l-1}>0)\right\}\right.\\ \times\Theta(x_1 > x_3 > \cdots > x_{2n-1})\Theta(x_{2n+1} > x_{2n+2} > \cdots > x_N)\\ \left.\times\,\delta(y_{2n+1})\dots\delta(y_N)\right),\end{aligned} \tag{3.94}$$

where

$$w_\nu(\hat{\mu}, z) = |z|^{\nu/2} e^{\eta_- z} g_\nu(\hat{\mu}, z)\,, \tag{3.95}$$

$$g_\nu(\hat{\mu}, z) = 2K_{\frac{\nu}{2}}(\eta_+|z|) \qquad \text{for } z\in\mathbb{R}\,, \tag{3.96}$$

$$\begin{aligned}[g_\nu(\hat{\mu}, z)]^2 &= [g_\nu(\hat{\mu}, z^*)]^2\\ &= 2\int_0^\infty \frac{dt}{t} e^{-2\eta_+^2 t(x^2-y^2)-\frac{1}{4t}} K_{\frac{\nu}{2}}\left(2\eta_+^2 t(x^2+y^2)\right)\\ &\quad\times \mathrm{erfc}(2\eta_+\sqrt{t}\,|y|) \qquad \text{for } z\in\mathbb{C}\,,\end{aligned} \tag{3.97}$$

$$c_N = (\hat{\mu}\text{-independent factor}) \times (2\hat{\mu})^{-N(N+\nu)}\eta_+^{-N(N+\nu-1)/2}\,. \tag{3.98}$$

Here $n$ and $N-2n$ denote the number of complex-conjugate pairs and the number of real eigenvalues, respectively. In (3.94), $\Theta(\cdots)$ is defined to be 1 when $\cdots$ is true and 0 otherwise. In what follows, we denote the expectation value with respect to $Z^{\mathrm{RMT}}_{N_f,\nu}(\hat{\mu},\{\tau_f\})$ by $\langle\,\rangle_{N_f}$.

The determinants can be absorbed into the measure $d\Omega_{N,\nu}$ with a replacement

$$w_\nu(\hat{\mu}, z) \;\to\; w_{N_f,\nu}(\hat{\mu}, z) \equiv w_\nu(\hat{\mu}, z) \prod_{f=1}^{N_f} (\tau_f^2 + z)\,. \tag{3.99}$$

Therefore, roughly speaking, the generalization of the quenched result [57, 58, 66] to $N_f > 0$ consists of replacing $w_\nu$ by $w_{N_f,\nu}$.

The spectral density is defined as usual. Because the measure (3.94) has $\delta$-peaks along the real axis, the spectral density consists of two parts:

$$R_{N_f,\nu}(\hat{\mu}; z) \equiv \Big\langle \sum_{i=1}^{N} \delta^2(z - z_i) \Big\rangle_{N_f} \tag{3.100}$$

$$= R^{\mathbb{C}}_{N_f,\nu}(\hat{\mu}; z) + \delta(y) R^{\mathbb{R}}_{N_f,\nu}(\hat{\mu}; x)\,, \qquad z \equiv x + iy\,. \tag{3.101}$$

$R^{\mathbb{C}}_{N_f,\nu}$ is the spectral density for complex eigenvalues $\in \mathbb{C}\backslash\mathbb{R}$, whereas $R^{\mathbb{R}}_{N_f,\nu}$ is that for exactly real eigenvalues $\in \mathbb{R}$. This two-piece structure is a distinctive feature of the (chiral) real Ginibre ensemble ($\beta = 1$) and is not seen in the other classes ($\beta = 2,\ 4$).

By definition, $R^{\mathbb{C}}_{N_f,\nu}(\mu; z)$ is obtained from (3.93) by tuning one of the eigenvalues to the given $z \in \mathbb{C}\backslash\mathbb{R}$. Then one of the other eigenvalues is automatically tuned to $z^*$. Without losing generality one can assume, say, $z_{N-1} \overset{!}{=} z$ and $z_N \overset{!}{=} z^*$. Then

$$\prod_{k=1}^{N} w_{N_f,\nu}(\hat{\mu}, z_k) \prod_{i<j}(z_i - z_j) = \prod_{k=1}^{N-2} w_{N_f,\nu}(\hat{\mu}, z_k) \prod_{i<j}^{N-2} (z_i - z_j)$$
$$\times\, w_{N_f,\nu}(\hat{\mu}, z) w_{N_f,\nu}(\hat{\mu}, z^*)(z - z^*) \prod_{\ell=1}^{N-2} (z_\ell - z)(z_\ell - z^*)\,. \tag{3.102}$$

Therefore, up to a multiplicative constant, we have

$$R^{\mathbb{C}}_{N_f,\nu}(\mu; z) \sim w_{N_f,\nu}(\hat{\mu}, z) w_{N_f,\nu}(\hat{\mu}, z^*)(z - z^*) \Big\langle \prod_{\ell=1}^{N-2} (z_\ell - z)(z_\ell - z^*) \Big\rangle_{N_f}. \tag{3.103}$$

Here comes an important observation: the characteristic polynomial on the RHS may be cast into the form

$$\Big\langle \prod_{\ell=1}^{N-2} (z_\ell - z)(z_\ell - z^*) \Big\rangle_{N_f} = \frac{1}{(\sqrt{-z}\sqrt{-z^*})^\nu} \Big\langle \det\Big[\sqrt{-z}\,\mathbf{1}_{2N+\nu} - \begin{pmatrix} \mathbf{0} & A \\ -B^T & \mathbf{0} \end{pmatrix}\Big]$$
$$\times \det\Big[\sqrt{-z^*}\,\mathbf{1}_{2N+\nu} - \begin{pmatrix} \mathbf{0} & A \\ -B^T & \mathbf{0} \end{pmatrix}\Big]\Big\rangle_{N_f}, \tag{3.104}$$

which is nothing but the partition function with $N_f+2$ flavors! Therefore

$$R^{\mathbb{C}}_{N_f,\nu}(\mu;z) \sim |z-z^*| \frac{w_{N_f,\nu}(\hat{\mu},z)w_{N_f,\nu}(\hat{\mu},z^*)}{|z|^\nu} \frac{Z^{\rm RMT}_{N_f+2,\nu}(\hat{\mu},\{\tau_f\})\Big|_{\tau_{N_f+1}=\sqrt{-z},\,\tau_{N_f+2}=\sqrt{-z^*}}}{Z^{\rm RMT}_{N_f,\nu}(\hat{\mu},\{\tau_f\})}. \tag{3.105}$$

By the same token we find for the real spectral density

$$R^{\mathbb{R}}_{N_f,\nu}(\mu;x) \sim \frac{w_{N_f,\nu}(\hat{\mu},x)}{\sqrt{-x}^{\,\nu}} \int_{\mathbb{R}} dt\, \frac{w_{N_f,\nu}(\hat{\mu},t)}{\sqrt{-t}^{\,\nu}} |x-t| \frac{Z^{\rm RMT}_{N_f+2,\nu}(\hat{\mu},\{\tau_f\})\Big|_{\tau_{N_f+1}=\sqrt{-x},\,\tau_{N_f+2}=\sqrt{-t}}}{Z^{\rm RMT}_{N_f,\nu}(\hat{\mu},\{\tau_f\})}. \tag{3.106}$$

We note that this structure of spectral densities, being proportional to the ratio of $Z_{N_f}$ and $Z_{N_f+2}$, is quite generic [68].

Thus, the spectral densities in the large-$N$ limit will follow if we could evaluate the asymptotic form of the partition functions for generic $N_f$.[16] In the next section we will present analytic formulae for those partition functions.

### 3.4.2 The Finite-Volume Partition Function

In this section we will provide a closed expression for the partition function, which will be used later in the derivation of the microscopic spectral density in the large-$N$ limits of weak and strong non-Hermiticity.

According to a remarkable mathematical theorem by Kieburg and Guhr [39], a partition function of a random matrix ensemble can be written as a pfaffian if its joint eigenvalue distribution function has a "factorization property" [39]. Since (3.93) does fulfill this condition, it follows that the partition function of our model can be written as a pfaffian. The explicit formula depends on the parity of $N_f$. For $\boxed{\text{even } N_f}$, we have

$$Z^{\rm RMT}_{N_f,\nu}(\hat{\mu},\{\tau_f\}) = C_{N,N_f,\nu}(\hat{\mu}) \frac{\underset{1\le i,j\le N_f}{\rm Pf}\left[(m_j^2-m_i^2)Z^{\rm RMT}_{2,\nu}(\hat{\mu},\{\tau_i,\tau_j\})\right]}{\Delta_{N_f}(\tau_1^2,\ldots,\tau_{N_f}^2)}, \tag{3.107}$$

with $C_{N,N_f,\nu}(\hat{\mu})$ a mass-independent factor and $\Delta_{N_f}(\tau_1^2,\ldots,\tau_{N_f}^2) \equiv \prod_{i>j}^{N_f}(\tau_i^2-\tau_j^2)$.[17] Since we are mainly interested in the mass dependence of $Z^{\rm RMT}_{N_f,\nu}$, we may simply write (3.107) as

[16] For finite $N$, the massive partition functions were obtained in [66] with the aid of skew-orthogonal polynomials. However we do not consider finite $N$, since the universal correspondence of ChRMT to QCD emerges only in the large-$N$ limit.

[17] It should not be confused with the BCS gap $\Delta$.

$$Z_{N_f,\nu}^{\mathrm{RMT}}(\hat{\mu}, \{\tau_f\}) \sim \frac{\mathop{\mathrm{Pf}}\limits_{1\le i,j\le N_f}\left[(\tau_j^2 - \tau_i^2) Z_{2,\nu}^{\mathrm{RMT}}(\hat{\mu}, \{\tau_i, \tau_j\})\right]}{\Delta_{N_f}(\tau_1^2, \ldots, \tau_{N_f}^2)}. \tag{3.108}$$

Note that this formula is trivially satisfied for $N_f = 2$.

The partition function for $\boxed{\text{odd } N_f}$ reads [39]

$$Z_{N_f,\nu}^{\mathrm{RMT}}(\hat{\mu}, \{\tau_f\}) \sim \frac{\mathrm{Pf}\begin{bmatrix} 0 & \left\{Z_{1,\nu}^{\mathrm{RMT}}(\hat{\mu}, \tau_j)\right\}_{1\le j\le N_f} \\ \left\{-Z_{1,\nu}^{\mathrm{RMT}}(\hat{\mu}, \tau_i)\right\}_{1\le i\le N_f} & \left\{(\tau_j^2 - \tau_i^2) Z_{2,\nu}^{\mathrm{RMT}}(\hat{\mu}, \{\tau_i, \tau_j\})\right\}_{1\le i,j\le N_f} \end{bmatrix}}{\Delta_{N_f}(\tau_1^2, \ldots, \tau_{N_f}^2)}. \tag{3.109}$$

Therefore the limiting form of the partition functions for generic $N_f$ at large $N$ will follow if it is known for $N_f = 1$ and $2$. Let us define

$$\mathcal{Z}_w^{(N_f,\nu)}(\gamma, \{\kappa_f\}) \equiv \lim_{N\to\infty,\ \mathrm{weak}} Z_{N_f,\nu}^{\mathrm{RMT}}(\hat{\mu}, \{\tau_f\}) \tag{3.110}$$

$$\mathcal{Z}_s^{(N_f,\nu)}(\{\tau_f\}) \equiv \lim_{N\to\infty,\ \mathrm{maximal}} Z_{N_f,\nu}^{\mathrm{RMT}}(\hat{\mu}, \{\tau_f\}) \tag{3.111}$$

where each limit stands for the large-$N$ limit of weak and maximal non-Hermiticity, respectively.

### Large-$N$ Limit at Maximal Non-Hermiticity: $\hat{\mu} = 1$, $\tau_f \sim \mathcal{O}(1)$ and $\nu = 0$

Considering a physical application to dense two-color QCD, it is enough to consider even $N_f$ and $\nu = 0$. At large $N$ the partition function reduces to (3.77), and especially

$$\mathcal{Z}_s^{(N_f=2,0)}(\tau_1, \tau_2) \sim I_0(\tau_1 \tau_2). \tag{3.112}$$

Substituting this into (3.108), we find

$$Z_s^{(N_f,0)}(\{\tau_f\}) \sim \frac{\mathop{\mathrm{Pf}}\limits_{1\le i,j\le N_f}\left[(\tau_j^2 - \tau_i^2) I_0(\tau_i \tau_j)\right]}{\Delta_{N_f}(\tau_1^2, \ldots, \tau_{N_f}^2)}. \tag{3.113}$$

**Large-$N$ Limit at Weak Non-Hermiticity: $N\hat{\mu}^2 \sim \sqrt{N}\tau_f \sim \mathcal{O}(1)$**

Let us define rescaled variables $\boxed{\kappa_f \equiv 2\sqrt{N}\tau_f}$ and $\boxed{\gamma \equiv \sqrt{2N}\hat{\mu}}$. In the large-$N$ limit of weak non-Hermiticity, (3.79) reduces to (3.83) for arbitrary $N_f$. For $N_f = 1$ the $\hat{\mu}$-dependent part factorizes ($\because\ UI_2U^T = I_2$ for $U \in \mathrm{SU}(2)$), so that

$$Z_w^{(N_f=1,\nu)}(\gamma, \kappa) \sim \mathrm{e}^{-\gamma^2/2} I_\nu(\kappa)\,. \tag{3.114}$$

While it is rather difficult to evaluate the coset integral for $Z_w^{(N_f=2,\nu)}(\gamma, \{\kappa_1, \kappa_2\})$ directly, we can exploit the weak kernel in [[58], Eq. (2.32)], to derive

$$Z_w^{(N_f=2,\nu)}(\gamma, \{\kappa_1, \kappa_2\}) \sim \frac{1}{\kappa_1^2 - \kappa_2^2} \int_0^1 ds\, s^2 \mathrm{e}^{-2\gamma^2 s^2} \Big\{ \kappa_1 I_{\nu+1}(s\kappa_1) I_\nu(s\kappa_2) - (\kappa_1 \leftrightarrow \kappa_2) \Big\}\,. \tag{3.115}$$

Plugging (3.114) and (3.115) into (3.108) and (3.109), the partition function for arbitrary $N_f$ can be obtained. It can be verified that in the limit $\gamma \to 0$ known formulae at $\hat{\mu} = 0$ [69] are reproduced, as expected.

### 3.4.3 Microscopic Spectral Density at Maximal Non-Hermiticity: $\hat{\mu} = 1$

**Main Results**

In this limit we take $N$ to infinity with $\{\tau_f\}$ and $\{\Lambda_k\}$ kept finite. Namely we do not have to rescale the eigenvalues for the Dirac operator (3.91).

Using (3.113) in (3.105) and (3.106), we immediately obtain the microscopic spectral density for complex and real eigenvalues:

$$\rho_s^{(N_f,\mathbb{C})}(z) \equiv \lim_{N\to\infty} R^{\mathbb{C}}_{N_f,0}(1; z) \tag{3.116}$$

$$= \frac{1}{32\pi} |z - z^*| w_{N_f,0}(1, z) w_{N_f,0}(1, z^*) H_s^{N_f}(\sqrt{-z}, \sqrt{-z^*})\,, \tag{3.117}$$

$$\rho_s^{(N_f,\mathbb{R})}(x) \equiv \lim_{N\to\infty} R^{\mathbb{R}}_{N_f,0}(1; x) \tag{3.118}$$

$$= \frac{1}{64\pi} \frac{w_{N_f,0}(1, x)}{\sqrt{-x}^{\nu}} \int_{\mathbb{R}} dt\, \frac{w_{N_f,0}(1, t)}{\sqrt{-t}^{\nu}} |x - t| H_s^{N_f}(\sqrt{-x}, \sqrt{-t})\,, \tag{3.119}$$

where we have defined

$$H_s^{N_f}(\alpha,\beta)=\frac{\frac{1}{\Delta_{N_f+2}(\tau_1^2,\dots,\tau_{N_f}^2,\alpha^2,\beta^2)}\underset{1\le i,j\le N_f+2}{\mathrm{Pf}}\left[(\tau_j^2-\tau_i^2)I_0(\tau_i\tau_j)\right]\Big|_{\tau_{N_f+1}=\alpha,\ \tau_{N_f+2}=\beta}}{\frac{1}{\Delta_{N_f}(\tau_1^2,\dots,\tau_{N_f}^2)}\underset{1\le i,j\le N_f}{\mathrm{Pf}}\left[(\tau_j^2-\tau_i^2)I_0(\tau_i\tau_j)\right]}\,. \tag{3.120}$$

The normalization constants are fixed through matching with the quenched result [58]; namely we impose the condition that $\rho_{s,\mathrm{Dirac}}^{(N_f,\mathbb{C})}(\Lambda)\to\frac{1}{\pi}$ as $|\Lambda|\to\infty$.

The microscopic spectral densities for Dirac eigenvalues are then obtained via (3.92) as

$$\rho_{s,\mathrm{Dirac}}^{(N_f,\mathbb{C})}(\Lambda)=4|\Lambda|^2\rho_s^{(N_f,\mathbb{C})}(-\Lambda^2)\qquad\text{for }\Lambda^2\in\mathbb{C}\setminus\mathbb{R}\,, \tag{3.121}$$

$$\rho_{s,\mathrm{Dirac}}^{(N_f,\mathbb{R})}(\Lambda)=2|\Lambda|\rho_s^{(N_f,\mathbb{R})}(-\Lambda^2)\qquad\text{for }\Lambda^2\in\mathbb{R}\,. \tag{3.122}$$

Note that both $\Lambda\in\mathbb{R}$ and $\Lambda\in i\mathbb{R}$ are included in the latter case. Due to the reality of the Dirac matrix,

$$\rho_{s,\mathrm{Dirac}}^{(N_f,\mathbb{C})}(\Lambda)=\rho_{s,\mathrm{Dirac}}^{(N_f,\mathbb{C})}(-\Lambda)=\rho_{s,\mathrm{Dirac}}^{(N_f,\mathbb{C})}(\Lambda^*)=\rho_{s,\mathrm{Dirac}}^{(N_f,\mathbb{C})}(-\Lambda^*)\,. \tag{3.123}$$

As a consistency check, let us confirm that the obtained spectral densities exhibit the decoupling of heavy flavors [41, 70]. Because $N_f$ is assumed to be even, one cannot decouple an odd number of flavors and at least two flavors must decouple. If $\tau_{N_f-1}$ and $\tau_{N_f}$ go to infinity, a bit of algebra shows that

$$H_s^{N_f}(\alpha,\beta)\to\frac{1}{(\tau_{N_f-1}\tau_{N_f})^4}H_s^{N_f-2}(\alpha,\beta)\,. \tag{3.124}$$

Using this, one can readily show the decoupling of the heavy two flavors:

$$\rho_{s,\mathrm{Dirac}}^{(N_f,\mathbb{C}/\mathbb{R})}(\Lambda)\to\rho_{s,\mathrm{Dirac}}^{(N_f-2,\mathbb{C}/\mathbb{R})}(\Lambda)\,. \tag{3.125}$$

This procedure can be iterated until $N_f=0$ is reached, and the final result coincides with the known result at $N_f=0$ [58], as it should.

In the chiral limit ($^\forall\tau_f=0$) there is a symmetry

$$\rho_{s,\mathrm{Dirac}}^{(N_f,\mathbb{C}/\mathbb{R})}(\Lambda)=\rho_{s,\mathrm{Dirac}}^{(N_f,\mathbb{C}/\mathbb{R})}(\pm i\Lambda)\,. \tag{3.126}$$

Actually this property has been conjectured on p. 81 to account for the vanishing of the spectral sum rules in the chiral limit. It is gratifying to see that it is indeed satisfied.

**Figures**

For $N_f = 2$ we plotted the microscopic spectral density of complex, real and purely imaginary eigenvalues in Figs. 3.1, 3.2 and 3.3. Because of the symmetry (3.123) it is enough to plot for the first quadrant ($\mathrm{Re}\Lambda > 0$, $\mathrm{Im}\Lambda > 0$). Owing to the Vandermonde determinant in (3.94), the complex eigenvalues repel each other, which in particular implies that $\Lambda \leftrightarrow \Lambda^*$ and $\Lambda \leftrightarrow -\Lambda^*$ repel each other, resulting in the vanishing of $\rho_{s,\mathrm{Dirac}}^{(N_f,\mathbb{C})}(\Lambda)$ along the real and imaginary axis.

For degenerate masses ($\tau_1 = \tau_2$) the fermion determinant is positive definite and so is the spectral density. The spectrum is smooth and closely resembles the result for $N_f = 0$ [58]. When we gradually increase $|\tau_1 - \tau_2|$ from zero, the sign problem emerges, and $\rho_{s,\mathrm{Dirac}}^{(N_f=2,\mathbb{C})}$ starts to show a rapid oscillation inside an ellipse with an enormous amplitude. (Were it not for the sign problem, the spectral density cannot take a negative value!) This phenomenon is quite analogous to what has been observed in three-color QCD at $V_4\mu^4 \ll 1$ [71–73] despite different symmetries. In Sect. 3.5.1 we will show how to understand the elliptical domain of oscillation in a simple way.

On the other hand, the spectral density on the real axis $\rho_{s,\mathrm{Dirac}}^{(N_f=2,\mathbb{R})}(x)$ looks more smooth. It changes sign twice for unequal masses, reflecting the indefinite sign of the measure. The spectral density on the imaginary axis $\rho_{s,\mathrm{Dirac}}^{(N_f=2,\mathbb{R})}(ix)$ is hardly affected by the mass parameters.

The figures obtained can naturally be looked upon as the superposition of the 'quenched' contribution and the 'unquenched', strongly oscillating contribution localized in an elliptical domain. Indeed $\rho_{s,\mathrm{Dirac}}^{(N_f,\mathbb{C})}(z)$ is, by definition, already written as a sum of such two pieces. To see this, one should expand the pfaffian in (3.117).

### *3.4.4 Microscopic Spectral Density at Weak Non-Hermiticity: $N\hat{\mu}^2 \sim \mathcal{O}(1)$*

**Main Results**

At weak non-Hermiticity, $N_f$ is not restricted to be even, and $\nu$ can take arbitrary integer values. In this limit we take $N$ to infinity while keeping $\boxed{\kappa_f = 2\sqrt{N}\tau_f}$ and $\boxed{\gamma = \sqrt{2N}\hat{\mu}}$ finite. The Dirac eigenvalues $\{\Lambda_k\}$ also have to be rescaled in the same way as $\{\tau_f\}$, namely $\boxed{\xi_k = 2\sqrt{N}\Lambda_k}$ is kept fixed.

Introducing the squared eigenvalues $s_k = -\xi_k^2$ we shall proceed in two steps as in the last section: we first derive the spectral density for $\{s_k\}$, and then convert it to the spectral density of $\{\xi_k\}$. With the new variables the fermion determinant in (3.93) reads

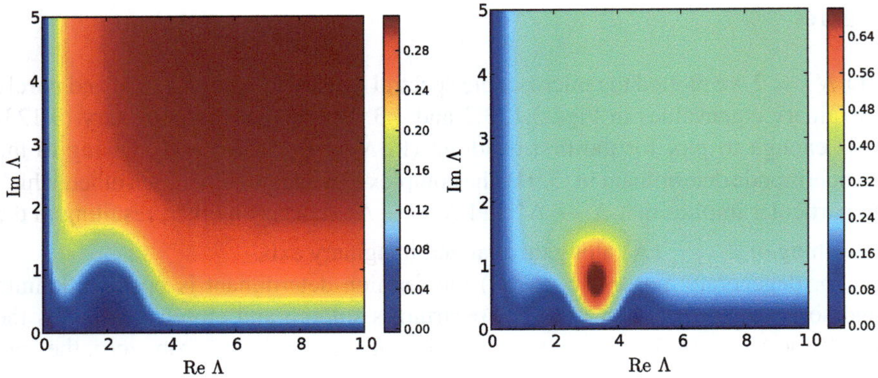


**Fig. 3.1** $\rho_{s,\mathrm{Dirac}}^{(N_f=2,\mathbb{C})}(\Lambda)$ for $(\tau_1, \tau_2) = (2,\ 2)$ (*left*) and $(2,\ 4.5)$ (*right*)

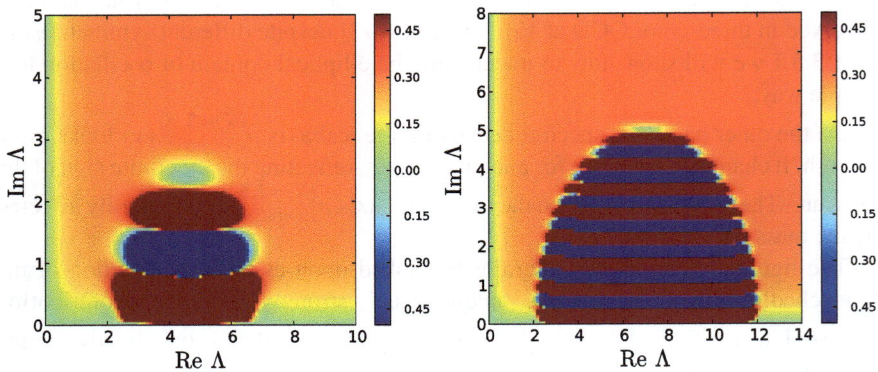


**Fig. 3.2** $\rho_{s,\mathrm{Dirac}}^{(N_f=2,\mathbb{C})}(\Lambda)$ for $(\tau_1, \tau_2) = (2,\ 7)$ (*left*) and $(2,\ 12)$ (*right*). (The values outside the range $[-0.5,\ 0.5]$ are clipped for a better visibility of the figure)

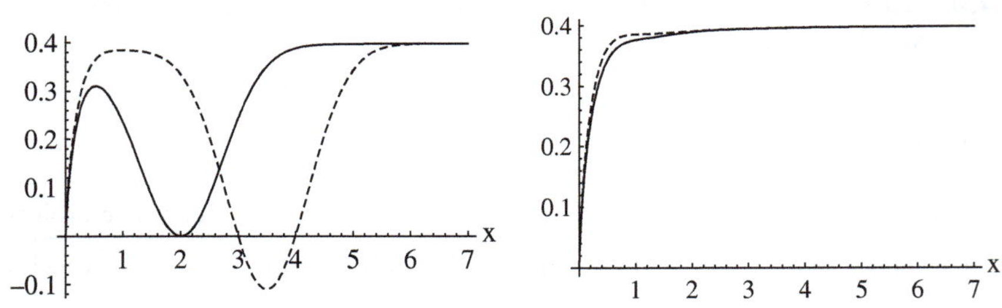


**Fig. 3.3** *Left:* $\rho_{s,\mathrm{Dirac}}^{(N_f=2,\mathbb{R})}(x)$ for $(\tau_1, \tau_2) = (2,\ 2)$ (*full line*) and $(3,\ 4)$(*dashed line*). *Right:* $\rho_{s,\mathrm{Dirac}}^{(N_f=2,\mathbb{R})}(ix)$ for $(\tau_1, \tau_2) = (2,\ 8)$ (*full line*) and $(3,\ 4)$ (*dashed line*)

$$\prod_{f=1}^{N_f}(\tau_f^2+z_k)=\prod_{f=1}^{N_f}\left(\frac{\kappa_f^2}{4N}+\frac{s_k}{4N}\right)\propto\prod_{f=1}^{N_f}(\kappa_f^2+s_k)\,. \tag{3.127}$$

The rescaled form of $w_\nu(\hat{\mu},z)$ is given in [58]. Therefore from (3.105) we obtain for $s\in\mathbb{C}\backslash\mathbb{R}$

$$\rho_w^{(N_f,\mathbb{C})}(\gamma;s)\equiv\lim_{N\to\infty}\frac{1}{(4N)^2}R^{\mathbb{C}}_{N_f,\nu}\Big(\frac{\gamma}{\sqrt{2N}};\frac{s}{4N}\Big)\qquad[s=p+iq] \tag{3.128}$$

$$=\frac{|q|}{64\pi\gamma^2}\mathrm{e}^{\frac{p}{4\gamma^2}}\mathcal{G}_\nu(\gamma,s)\prod_{f=1}^{N_f}|\kappa_f^2+s|^2H_w^{N_f,\nu}(\gamma;\sqrt{-s},\sqrt{-s^*})\,, \tag{3.129}$$

$$\mathcal{G}_\nu(\gamma,s)\equiv\lim_{N\to\infty}g_\nu\Big(\frac{\gamma}{\sqrt{2N}},\frac{s}{4N}\Big)^2\qquad\big(g_\nu\Rightarrow(3.97)\big) \tag{3.130}$$

$$=2\int_0^\infty\frac{dt}{t}\exp\Big[-\frac{(p^2-q^2)t}{32\gamma^4}-\frac{1}{4t}\Big]K_{\frac{\nu}{2}}\left(\frac{(p^2+q^2)t}{32\gamma^4}\right)\mathrm{erfc}\left(\frac{\sqrt{t}\,|q|}{4\gamma^2}\right). \tag{3.131}$$

The normalization constant was fixed through matching with the result for $N_f=0$ [58].

The new function $H_w^{N_f,\nu}(\gamma;\alpha,\beta)$ originates from (3.114) and (3.115) plugged into (3.108) and (3.109). It is defined as below:

- For even $N_f$:

$$H_w^{N_f,\nu}(\gamma;\alpha,\beta)$$

$$=\frac{\left.\frac{1}{\Delta_{N_f+2}(\kappa_1^2,\dots,\kappa_{N_f}^2,\alpha^2,\beta^2)}\underset{1\le i,j\le N_f+2}{\mathrm{Pf}}\Big[\mathfrak{H}_\nu(\gamma,\{\kappa_i,\kappa_j\})\Big]\right|_{\kappa_{N_f+1}=\alpha,\,\kappa_{N_f+2}=\beta}}{\frac{1}{\Delta_{N_f}(\kappa_1^2,\dots,\kappa_{N_f}^2)}\underset{1\le i,j\le N_f}{\mathrm{Pf}}\Big[\mathfrak{H}_\nu(\gamma,\{\kappa_i,\kappa_j\})\Big]}, \tag{3.132}$$

- For odd $N_f$:

$$
\begin{aligned}
&H_w^{N_f,\nu}(\gamma;\alpha,\beta)\\
&= \frac{\mathrm{Pf}\begin{bmatrix} 0 & \left\{I_\nu(\kappa_j)\right\}_{1\le j\le N_f+2} \\ \left\{-I_\nu(\kappa_i)\right\}_{1\le i\le N_f+2} & \left\{\mathfrak{H}_\nu(\gamma,\{\kappa_i,\kappa_j\})\right\}_{1\le i,j\le N_f+2}\end{bmatrix}_{\kappa_{N_f+1}=\alpha,\,\kappa_{N_f+2}=\beta}}{\Delta_{N_f+2}(\kappa_1^2,\ldots,\kappa_{N_f}^2,\alpha^2,\beta^2)}\\
&\times\left\{\frac{1}{\Delta_{N_f}(\kappa_1^2,\ldots,\kappa_{N_f}^2)}\mathrm{Pf}\begin{bmatrix} 0 & \left\{I_\nu(\kappa_j)\right\}_{1\le j\le N_f} \\ \left\{-I_\nu(\kappa_i)\right\}_{1\le i\le N_f} & \left\{\mathfrak{H}_\nu(\gamma,\{\kappa_i,\kappa_j\})\right\}_{1\le i,j\le N_f}\end{bmatrix}\right\}^{-1},
\end{aligned}
\tag{3.133}
$$

where (cf. (3.115))

$$
\mathfrak{H}_\nu(\gamma,\{\kappa_1,\kappa_2\}) \equiv -\int_0^1 du\, u^2 e^{-2\gamma^2u^2}\left\{\kappa_1 I_{\nu+1}(u\kappa_1)I_\nu(u\kappa_2)-(\kappa_1\leftrightarrow\kappa_2)\right\}. \tag{3.134}
$$

Finally, using $s_k=-\xi_k^2$, we arrive at the microscopic spectral density for $\{\xi_k\}$:

$$
\rho_{w,\mathrm{Dirac}}^{(N_f,\mathbb{C})}(\gamma;\xi) = 4|\xi|^2\rho_w^{(N_f,\mathbb{C})}(\gamma;-\xi^2) \quad \text{for } \xi^2\in\mathbb{C}\setminus\mathbb{R}\,. \tag{3.135}
$$

The microscopic density of exactly real and purely imaginary Dirac eigenvalues is much harder to compute. To get it one has to compute

$$
\rho_w^{(N_f,\mathbb{R})}(\gamma;x) \equiv \lim_{N\to\infty}\frac{1}{4N}R_{N_f,\nu}^{\mathbb{R}}\left(\frac{\gamma}{\sqrt{2N}};\frac{x}{4N}\right) \qquad \text{for } x\in\mathbb{R}\,, \tag{3.136}
$$

but it happens that, if we substitute the large-$N$ limit of the partition functions into (3.106), the integral does not converge. It means that the large-$N$ limit and the integration $\int_{\mathbb{R}} dt$ *do not commute.* (This problem does not occur in the strong non-Hermiticity). This intricacy is not due to the chemical potential because the same problem arises even at $\hat{\mu}=0$ [47]. Through a highly complicated calculation, $\rho_w^{(N_f,\mathbb{R})}(\gamma;x)$ and $\rho_{w,\mathrm{Dirac}}^{(N_f,\mathbb{R})}(\gamma;\xi)$ for $N_f=0$ and 1 have been obtained by M. J. Phillips [74, 75]. His result for $N_f=0$ reads

$$
\rho_{w,\mathrm{Dirac}}^{(N_f=0,\mathbb{R})}(\gamma;\xi) \equiv 2|\xi|\rho_w^{(N_f=0,\mathbb{R})}(\gamma;-\xi^2) \tag{3.137}
$$

$$
= -2|\xi|G_w(-\xi^2,-\xi^2) \qquad \text{for } \xi^2\in\mathbb{R}\,, \tag{3.138}
$$

where

$$G_w(x,x') \equiv -\frac{\hat{h}_w(x')}{[\,\mathrm{sgn}(x')]^{\nu/2}}\Bigg\{\left((-i)^\nu \int_{-\infty}^{0} dy + \frac{2}{[\,\mathrm{sgn}(x')]^{\nu/2}}\int_0^{x'} dy\right)\mathcal{K}_w(x,y)\hat{h}_w(y)$$
$$-\frac{1}{32\sqrt{\pi}}\Bigg[-\frac{1}{\gamma}\,\mathrm{e}^{-\gamma^2} J_\nu(\sqrt{x}) + \frac{2\gamma^\nu}{\Gamma\left(\frac{\nu+1}{2}\right)}\int_0^1 ds\,\mathrm{e}^{-\gamma^2 s^2} s^{\nu+2}$$
$$\times\left(\frac{\sqrt{x}}{2}E_{\frac{1-\nu}{2}}(\gamma^2 s^2)\,J_{\nu+1}(s\sqrt{x}) - \gamma^2 s\left(E_{\frac{-1-\nu}{2}}(\gamma^2 s^2) - E_{\frac{1-\nu}{2}}(\gamma^2 s^2)\right)J_\nu(s\sqrt{x})\right)\Bigg]\Bigg\}. \tag{3.139}$$

Here we have defined

$$\hat{h}_w(x) \equiv \mathrm{e}^{x/8\gamma^2}\, 2K_{\frac{\nu}{2}}\left(\frac{|x|}{8\gamma^2}\right), \tag{3.140}$$

$$E_n(x) \equiv \int_1^\infty dt\, t^{-n}\,\mathrm{e}^{-xt}, \tag{3.141}$$

$$\mathcal{K}_w(u,v) \equiv \frac{1}{256\pi\gamma^2}\int_0^1 ds\; s^2 \mathrm{e}^{-2\gamma^2 s^2}\left\{\sqrt{u}J_{\nu+1}(s\sqrt{u})J_\nu(s\sqrt{v}) - (u \leftrightarrow v)\right\}. \tag{3.142}$$

I skip the derivation because it is too extensive to be reproduced here.

In Sect. 3.5 we will use (3.138) and (3.139) to analyze the sign problem.

**Figures**

The microscopic spectral density of complex eigenvalues for $N_f = 1$ in the limit of weak non-Hermiticity is plotted in Figs. 3.4, 3.5, 3.6 and 3.7 for multiple values of the winding number ($\nu$), the chemical potential ($\gamma$) and the quark mass ($\kappa$). As we increase $\gamma$ a region of strong oscillation appears gradually, reflecting the sign problem in $N_f = 1$ theory. It is observed that the oscillation gets milder when $\kappa$ or $\nu$ is increased at fixed $\gamma$. This is consistent with the observation in [76] for three-color QCD. Qualitatively, the microscopic spectrum of two-color QCD at low density is quite similar to that of three-color QCD, except for the apparent distinction that the complex spectral density in two-color QCD has no imaginary part, and vanishes in the vicinity of the real and imaginary axis.

## 3.5 Sign Problem in Dense Two-Color QCD and ChRMT

### *3.5.1 Structure of the Dirac Spectrum*

As we have seen in the previous section, for certain ranges of the parameters the unquenched microscopic spectral density of complex eigenvalues consists of

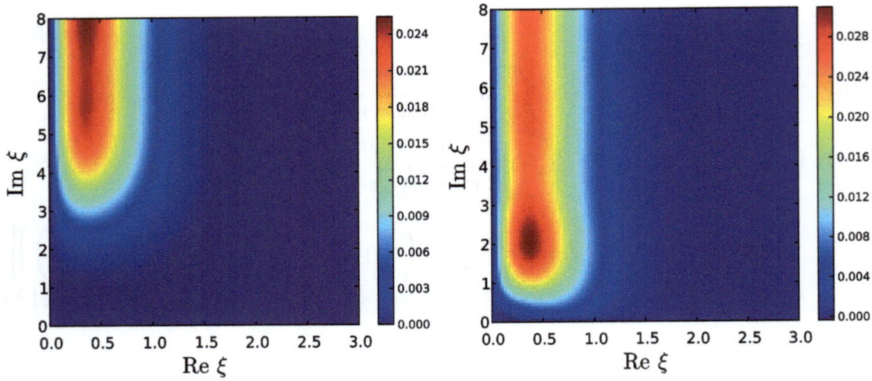


**Fig. 3.4** $\rho_{w,\mathrm{Dirac}}^{(N_f=1,\mathbb{C})}(\sqrt{0.2};\xi)$ with $\nu=0$ for $\kappa=0$ (*left*) and for $\kappa\to\infty$ (*right*), i.e., the quenched limit

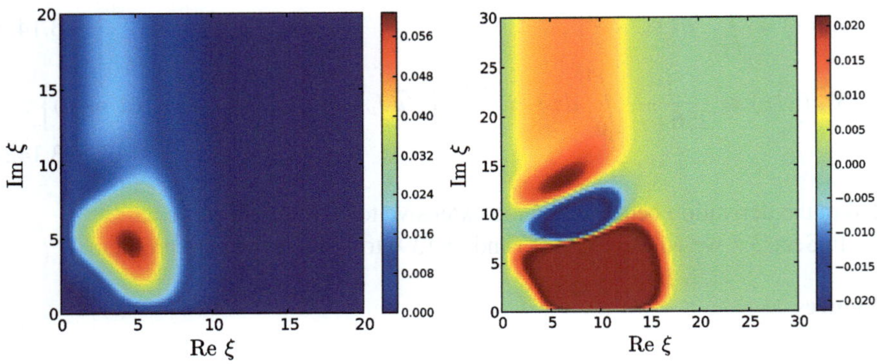


**Fig. 3.5** $\rho_{w,\mathrm{Dirac}}^{(N_f=1,\mathbb{C})}(1.8;\xi)$ (**left**) and $\rho_{w,\mathrm{Dirac}}^{(N_f=1,\mathbb{C})}(2.5;\xi)$ (*right*) with $\nu=0$ and $\kappa=0$. (In the *right figure*, the values outside the range $[-0.02,\ 0.02]$ are clipped for a better visibility of the figure)

a smooth flat part plus a circular domain of rapid oscillation with large amplitude. At strong non-Hermiticity with $N_f=2$, this oscillation occurs when quark masses are not degenerate. At weak non-Hermiticity with $N_f=1$, the oscillation grows as we increase $\mu$. These are exactly the cases afflicted with a severe sign problem. Indeed, were it not for the sign problem the spectral density must be positive definite and an oscillation between positive and negative values cannot take place. Therefore this behavior can be seen as a barometer of the sign problem in the underlying theory.

In this subsection we present a simple method to deduce the oscillatory behavior from the exact formula of the spectral density. Although we will focus on the case $N_f=2$ at maximal non-Hermiticity (corresponding to two-color QCD in the high density BCS phase) for simplicity, our methods are by no means limited to this particular case.

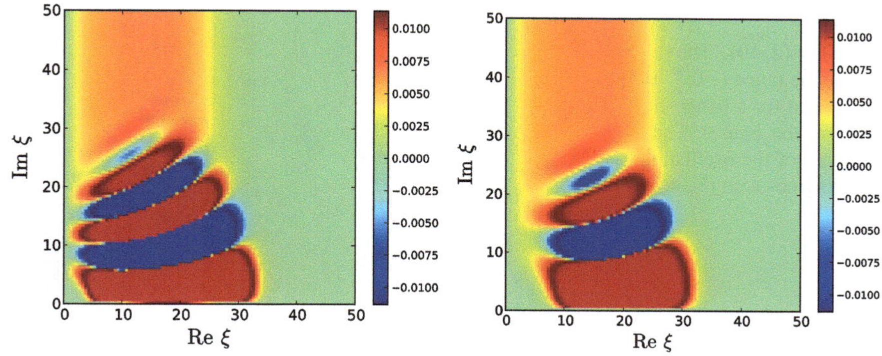


**Fig. 3.6** $\rho_{w,\mathrm{Dirac}}^{(N_f=1,\mathbb{C})}(3.5;\xi)$ with $\kappa = 0$ for $\nu = 0$ (*left*) and $\nu = 3$ (*right*). (The values outside the range $[-0.01,\ 0.01]$ are clipped for a better visibility of the figure)

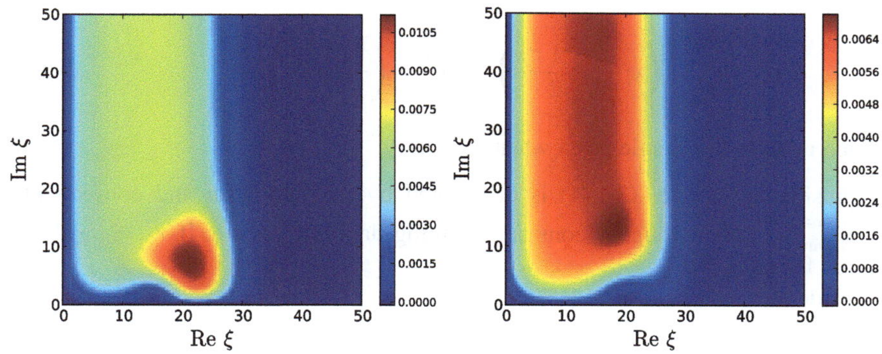


**Fig. 3.7** $\rho_{w,\mathrm{Dirac}}^{(N_f=1,\mathbb{C})}(3.5;\xi)$ with $\nu = 0$ for $\kappa = 15$ (*left*) and $\kappa = 20$ (*right*)

First, we assume $\boxed{1 \ll \tau_f \text{ and } 1 \ll \mathrm{Re}\,\Lambda}$ to simplify the exact expression (3.121) for $\rho_{s,\mathrm{Dirac}}^{(N_f=2,\mathbb{C})}(\Lambda)$. In this limit the $\varepsilon$-regime is continuously connected to the $p$-regime. For such large parameters one can replace the modified Bessel functions in (3.117) and (3.120) by their asymptotic forms. Expanding the pfaffian of (3.120) and only retaining the leading exponential factors, we obtain

$$\rho_{s,\mathrm{Dirac}}^{(N_f=2,\mathbb{C})}(\Lambda) \sim \mathrm{e}^{-|\Lambda|^2-\tau_1\tau_2}\left(\mathrm{e}^{|\Lambda|^2+\tau_1\tau_2} + \mathrm{e}^{\tau_1\Lambda+\tau_2\Lambda^*} + \mathrm{e}^{\tau_1\Lambda^*+\tau_2\Lambda}\right) \tag{3.143}$$

$$= 1 + \mathrm{e}^{-|\Lambda|^2-\tau_1\tau_2+\tau_1\Lambda+\tau_2\Lambda^*} + \mathrm{e}^{-|\Lambda|^2-\tau_1\tau_2+\tau_1\Lambda^*+\tau_2\Lambda}\,. \tag{3.144}$$

If $\mathrm{Re}(-|\Lambda|^2 - \tau_1\tau_2 + \tau_1\Lambda + \tau_2\Lambda^*) > 0$, the second and the third terms are exponentially greater than 1. Conversely, if $\mathrm{Re}(-|\Lambda|^2 - \tau_1\tau_2 + \tau_1\Lambda + \tau_2\Lambda^*) < 0$, they are exponentially suppressed. The boundary between the two domains is given by

$$\{\Lambda \in \mathbb{C} \mid \mathrm{Re}(-|\Lambda|^2 - \tau_1\tau_2 + \tau_1\Lambda + \tau_2\Lambda^*) = 0\}\,. \tag{3.145}$$

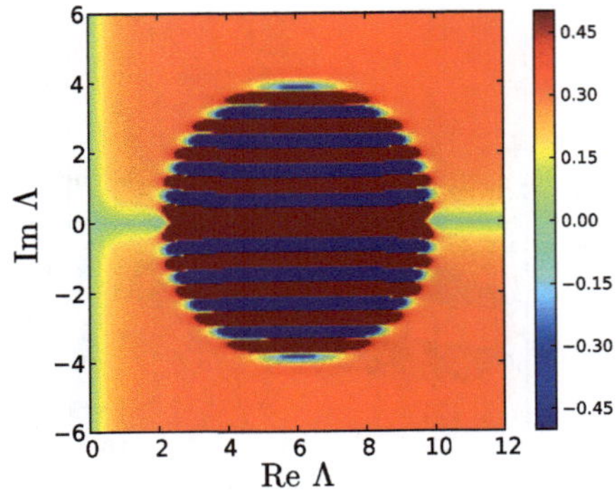


**Fig. 3.8** $\rho_{s,\mathrm{Dirac}}^{(N_f=2,\mathbb{C})}(\Lambda)$ with $(\tau_1, \tau_2) = (2, 10)$. (The values outside the range $[-0.5,\ 0.5]$ are clipped for a better visibility of the figure). The boundary of the oscillating region agrees with (3.146)

With $\Lambda \equiv x + iy$ we find that this set is nothing but the circle

$$\left(x - \frac{\tau_1 + \tau_2}{2}\right)^2 + y^2 = \left(\frac{\tau_1 - \tau_2}{2}\right)^2, \tag{3.146}$$

which reduces to a single point when $\tau_1 = \tau_2$.

- Outside this circle, (3.144) is dominated by the first term and consequently $\rho_{s,\mathrm{Dirac}}^{(N_f=2,\mathbb{C})}(\Lambda)$ is roughly constant; Nothing dramatic occurs. If $\tau_1 = \tau_2$ this is true for all $\Lambda \in \mathbb{C}$.
- Inside the circle,

$$\rho_{s,\mathrm{Dirac}}^{(N_f=2,\mathbb{C})}(\Lambda) \sim \mathrm{e}^{-|\Lambda|^2 - \tau_1\tau_2 + \tau_1\Lambda + \tau_2\Lambda^*} + \text{c.c.} \tag{3.147}$$

$$= 2\exp\left[\left(\frac{\tau_1 - \tau_2}{2}\right)^2 - \left(x - \frac{\tau_1 + \tau_2}{2}\right)^2 - y^2\right]\cos[(\tau_1 - \tau_2)y]\,. \tag{3.148}$$

This is an oscillatory function. The amplitude is $\mathcal{O}(1)$ at the circumference and grows larger in the interior of the circle, taking the maximal value $\exp\left[\left(\frac{\tau_1-\tau_2}{2}\right)^2\right]$ at the center of the circle. The ripplets are parallel to the $x$ axis and its wavelength along the $y$ axis is given by $\frac{2\pi}{|\tau_1-\tau_2|}$. The number of peaks of the wave inside the circle, denoted as $N_{\mathrm{peaks}}$, is roughly equal to

$$N_{\mathrm{peaks}} \approx |\tau_1 - \tau_2| \Big/ \frac{2\pi}{|\tau_1 - \tau_2|} \tag{3.149}$$

$$= \frac{|\tau_1 - \tau_2|^2}{2\pi}\,. \tag{3.150}$$

Now we compare our analytical estimates with the exact plot of $\rho_{s,\mathrm{Dirac}}^{(N_f=2,\mathbb{C})}(\Lambda)$ given in Fig. 3.8. The masses are set to $(\tau_1, \tau_2) = (2, 10)$. (In this figure the plot is cut off by hand for a better look of the plot). The sharp boundary between the oscillating region and the flat smooth region in Fig. 3.8 shows good agreement with our analytical estimate (3.146): $(x-6)^2+y^2 = 4^2$ for this case. Moreover Fig. 3.8 shows roughly 10 waves inside the circle, in good agreement with our prediction $N_{\text{peaks}} \approx 64/2\pi \simeq 10.2$. Thus our leading-order analytical estimate gives an accurate account of the global structure of the spectral density, even for masses that do not satisfy $\tau_f \gg 1$.

As we know (cf. (3.105)) the complex spectral density $\rho_{s,\text{Dirac}}^{(N_f,\mathbb{C})}(\Lambda)$ is proportional to the partition function with *two additional flavors* $\mathcal{Z}_s^{(N_f+2)}(\{\tau_f\}, \Lambda, \Lambda^*)$, and this is where the factor $\mathrm{e}^{|\Lambda|^2+\tau_1\tau_2} + \mathrm{e}^{\tau_1\Lambda+\tau_2\Lambda^*} + \mathrm{e}^{\tau_1\Lambda^*+\tau_2\Lambda}$ in (3.143) originates from. These terms express three patterns of the diquark pairing, namely $\{(1,2),(3,4)\}$, $\{(1,3),(2,4)\}$ and $\{(1,4),(2,3)\}$, respectively. The $\begin{cases}\text{first} \\ \text{second and third}\end{cases}$ pairings are dominant in the ground state when $\Lambda$ is $\begin{cases}\text{inside} \\ \text{outside}\end{cases}$ the circle (3.146). Therefore the envelope of the oscillatory domain on the microscopic spectrum is nothing but the line of phase transition in $\mathcal{Z}_s^{(N_f+2)}(\{\tau_f\}, \Lambda, \Lambda^*)$!

The extension of this analysis to $N_f \geq 4$ at maximal non-Hermiticity will be completely straightforward. On the other hand, because of the additional parameter $\gamma = \sqrt{2N}\hat{\mu}$ the extension to the limit of weak non-Hermiticity will be less trivial.

### 3.5.2 Average Sign Factor

#### Introductory Remarks

In this subsection we study the severity of the sign problem by looking at the average sign of the Dirac determinant. The definition of such an average is not unique, and we will use the following guiding principles to define an average that behaves reasonably:

- The average should be bounded by 1 from above.
- Starting at 1 for $\hat{\mu} = 0$ (or degenerate masses) the average should be a decreasing function of $\mu$ (or the mass difference).

Let us illustrate the problem of finding a proper definition using the case of three-color QCD, which has been well studied in the RMT framework [76–80]. If we denote the complex phase of $\det[\mathcal{D}(\mu)+m]$ by $\mathrm{e}^{i\theta}$, the authors proposed to compute

$$\left\langle \mathrm{e}^{2i\theta} \right\rangle_{N_f}^{\text{QCD3}} \equiv \left\langle \frac{\det[\mathcal{D}(\mu)+m]}{\det[\mathcal{D}(\mu)+m]^\dagger} \right\rangle_{N_f}^{\text{QCD3}}, \tag{3.151}$$

which gives the expectation value of twice the phase in the case that the determinant is complex. Had we chosen the opposite ratio on the right-hand side instead, this would naively have led to $\langle e^{-2i\theta}\rangle_{N_f}$. However, for $N_f = 1$ we have in three-color QCD

$$\left\langle \frac{\det[\mathcal{D}+m]^\dagger}{\det[\mathcal{D}+m]} \right\rangle^{\text{QCD3}}_{N_f=1} = \frac{\langle \det[\mathcal{D}+m]^\dagger \rangle^{\text{QCD3}}_{N_f=0}}{\langle \det[\mathcal{D}+m] \rangle^{\text{QCD3}}_{N_f=0}} = 1\,, \tag{3.152}$$

because the expectation value in denominator and numerator (also called characteristic polynomial) is equal and $\mu$-independent [81]. Obviously the average sign factor defined this way fails to capture the sign problem. The lesson learned from here is that *using expectation values w.r.t. a measure which is not positive definite is fallacious because they do not allow for a probabilistic interpretation.*

In two-color QCD, the determinant has no complex phase but only sign. An ambiguity exists regarding how to define the average sign factor. Let us see a possible *bad choice* for the average sign. Due to the obvious inequality

$$\big\langle\, |\det[\mathcal{D}+m]|\, \big\rangle_{N_f=0} \geq \big\langle \det[\mathcal{D}+m] \big\rangle_{N_f=0} \tag{3.153}$$

we have

$$\langle\, \text{sgn} \det[\mathcal{D}+m] \rangle_{N_f=1} = \left\langle \frac{|\det[\mathcal{D}+m]|}{\det[\mathcal{D}+m]} \right\rangle_{N_f=1} \geq 1\,, \tag{3.154}$$

and hence a ratio of this kind, which would have been the natural generalization of (3.151), can be ruled out.

To guarantee that such an unphysical behavior does not occur, we should employ a positive definite measure. Our definition of the average sign for two-color QCD, which we will use throughout this section, is

$$p^{(N_f)}(\hat\mu; \{\tau_f\}) \equiv \left\langle \text{sgn} \prod_{f=1}^{N_f} \det[\mathcal{D}(\hat\mu) + \tau_f] \right\rangle_{||N_f||}\,, \tag{3.155}$$

where the average is computed in a sign-quenched theory denoted by $\|N_f\|$ (i.e., only the absolute values of the determinants appear in the measure). We recall that in the strong limit we always need $N_f$ to be even. Below we will consider two particular cases, the limit of weak non-Hermiticity with one flavor, where

$$p_w^{(N_f=1)}(\gamma; \kappa) = \lim_{N\to\infty,\ \text{weak}} p^{(N_f=1)}(\hat\mu; \tau)\,, \tag{3.156}$$

with $\boxed{\kappa = 2\sqrt{N}\tau}$ and $\boxed{\gamma \equiv \sqrt{2N}\hat\mu}$ fixed (cf. p. 94), and the limit of maximal non-Hermiticity with two flavors, where

$$p_s^{(N_f=2)}(\tau_1, \tau_2) = \lim_{N\to\infty,\ \text{maximal}} p^{(N_f=2)}(\hat\mu; \{\tau_1, \tau_2\})\,, \tag{3.157}$$

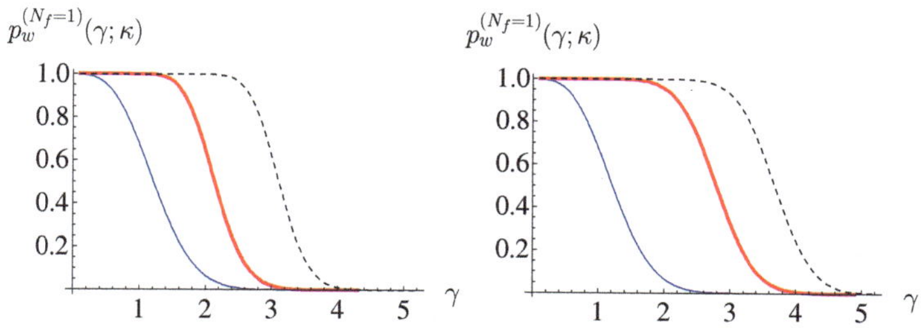


**Fig. 3.9** Average sign for $N_f = 1$ at weak non-Hermiticity. *Left:* $\kappa = 0$ (*blue line*), $\kappa = 6$ (*red thick line*) and $\kappa = 15$ (*black dashed line*), all for $\nu = 0$. *Right:* $\nu = 0$ (*blue line*), $\nu = 10$ (*red thick line*) and $\nu = 20$ (*black dashed line*), all for $\kappa = 0$

with $\tau_f$ fixed and $\hat{\mu} = 1$.

### Average Sign at Low Density

Let us consider the case $N_f = 1$ in the limit of weak non-Hermiticity. Although the sign-quenched partition function does not permit a field-theoretical interpretation like QCD with nonzero isospin chemical potential, it is still possible to compute it as a mathematical entity by RMT. In the Appendix A.4.3 we derive

$$\begin{aligned} \mathcal{Z}_w^{(N_f=\|1\|,\,\nu)}(\gamma,\kappa) &\equiv \lim_{N\to\infty,\ \text{weak}} \left\langle \left| \det[\mathcal{D}(\hat{\mu})+\tau] \right| \right\rangle_{N_f=0} \\ &\sim -\gamma\, \mathrm{e}^{\gamma^2/2} \frac{G_w(-\kappa^2,-\kappa^2)}{\hat{h}_w(-\kappa^2)}, \end{aligned} \tag{3.158}$$

where $G_w$ and $\hat{h}_w$ are defined in (3.139) and (3.140), respectively. This result essentially follows from the quenched real spectral density. Combining it with (3.114),

$$\mathcal{Z}_w^{(N_f=1,\,\nu)}(\gamma,\kappa) \sim \mathrm{e}^{-\gamma^2/2} I_\nu(\kappa)$$

we obtain the average sign factor in the weak limit,

$$p_w^{(N_f=1)}(\gamma;\kappa) = \frac{\mathcal{Z}_w^{(N_f=1,\,\nu)}(\gamma,\kappa)}{\mathcal{Z}_w^{(N_f=\|1\|,\,\nu)}(\gamma,\kappa)} \sim -\frac{\mathrm{e}^{-\gamma^2} I_\nu(\kappa)\hat{h}_w(-\kappa^2)}{\gamma\, G_w(-\kappa^2,-\kappa^2)}. \tag{3.159}$$

The normalization of $p_w^{(N_f=1)}(\gamma;\kappa)$ can be uniquely determined from $p_w^{(N_f=1)}(0;\kappa) = 1$.

In Fig. 3.9 we depict how $p_w^{(N_f=1)}(\gamma;\kappa)$ depends on $\gamma$ for several values of $\kappa$ and $\nu$. We observe that the onset of the sign problem is delayed as $\kappa$ or $\nu$ is increased, a feature which is in common with three-color QCD [77, 76]. In Fig. 3.10 we plot

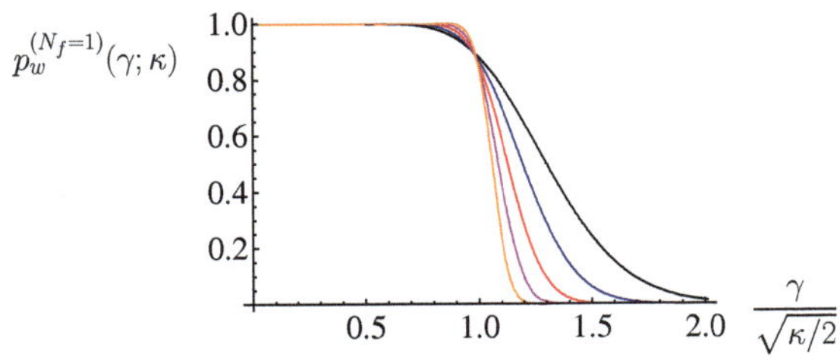


**Fig. 3.10** Average sign for $N_f = 1$ and $\nu = 0$ at weak non-Hermiticity. The curves correspond to $\kappa = 4$ (*black*), $\kappa = 8$ (*blue*), $\kappa = 16$ (*red*), $\kappa = 32$ (*magenta*), and $\kappa = 64$ (*orange*)

the average sign against a rescaled variable $\gamma/\sqrt{\kappa/2}$. In physical units, it is equal to $\mu/(m_\pi/2)$, where $m_\pi$ denotes the mass of the Nambu–Goldstone (NG) bosons in the vacuum.[18] Interestingly, the sign problem looks almost absent for $\mu < m_\pi/2$ while it deteriorates rapidly for $\mu > m_\pi/2$. In particular, we observe a convergence of the curves to a step function in the thermodynamic limit. This apparent jump of the average sign is quite intriguing, compared to the average phase factor in the microscopic limit of three-color QCD [77, 78]: the latter changes smoothly from 1 to 0 in the limit $\kappa \to \infty$.

Since the $\gamma$-dependence of $\mathcal{Z}_w^{(N_f=1,\,\nu)}$ is explicit in (3.114) (being of oscillatory nature), it must be $\mathcal{Z}_w^{(N_f=\|1\|,\,\nu)}$ that is responsible for the apparent jump of the average sign. While it seems that the $N_f = \|1\|$ theory is similar to the $N_f \geq 2$ theory in that both undergo a transition at $\mu = m_\pi/2$, a comparison of the order of the transition needs further investigation.

Finally we comment on a lattice simulation using one flavor of staggered fermions in the adjoint representation of SU(2) [82, 83]. This theory belongs to the same symmetry class as $N_f = 1$ two-color QCD with fundamental fermions. In [82, 83] this theory was simulated by two different algorithms, Hybrid-Monte-Carlo (HMC) and Two-Step-Multi-Boson (TSMB). The HMC algorithm is non-ergodic in this case and actually simulates a different theory, namely the one-flavor sign-quenched theory. The authors of [82, 83] report that the sign-quenched theory simulated by HMC seems to undergo a phase transition at $\mu = m_\pi/2$, while the correct one-flavor theory simulated by TSMB exhibits no singularity at all. These results are consistent with ours. However, the observables considered here and in [82, 83] are different so that a direct comparison is difficult.

[18] Note that no NG mode appears in $N_f = 1$ two-color QCD. The above $m_\pi$ refers to the mass of the NG modes that appear in $N_f \geq 2$ two-color QCD.

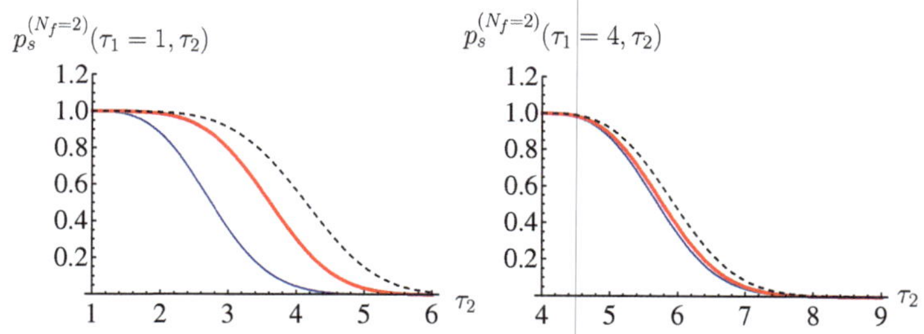


**Fig. 3.11** Average sign factor for $N_f = 2$ at maximal non-Hermiticity ($\hat{\mu} = 1$) with $\tau_1 = 1$ (*left*) and $\tau_1 = 4$ (*right*), at $\nu = 0$ (*blue line*), $\nu = 10$ (*red thick line*), and $\nu = 20$ (*black dashed line*)

### Average Sign at High Density

Let us now consider the case $N_f = 2$ in the limit of maximal non-Hermiticity ($\hat{\mu} = 1$). This theory is free from the sign problem if and only if the masses are degenerate. Here we are interested in the severity of the sign problem for $|\tau_1 - \tau_2|$ nonzero. In the Appendix A4.4 we derive

$$\begin{aligned}\mathcal{Z}_s^{(N_f=\|2\|,\,\nu)}(\tau_1, \tau_2) &\equiv \lim_{N\to\infty,\ \text{maximal}} \Big\langle \big| \det[\mathcal{D}(\hat{\mu}) + \tau_1] \det[\mathcal{D}(\hat{\mu}) + \tau_2] \big| \Big\rangle_{N_f=0} \\ &\sim \frac{|\tau_1 \tau_2|^{\nu}}{\left|\tau_1^2 - \tau_2^2\right|} f_{\infty}(-\tau_1^2, -\tau_2^2)\,, \qquad (3.160)\end{aligned}$$

where $f_\infty$ is defined in the Appendix A.4.4. This result essentially follows from the two-point correlation function of real squared Dirac eigenvalues. Combining (3.160) with (3.113) for $N_f = 2$,

$$\mathcal{Z}_s^{(N_f=2,\,\nu)}(\tau_1, \tau_2) \sim I_0(\tau_1 \tau_2)\,,$$

we obtain

$$p_s^{(N_f=2)}(\tau_1, \tau_2) = \frac{\mathcal{Z}_s^{(N_f=2,\,\nu)}(\tau_1, \tau_2)}{\mathcal{Z}_s^{(N_f=\|2\|,\,\nu)}(\tau_1, \tau_2)} = \frac{|\tau_1^2 - \tau_2^2|}{|\tau_1 \tau_2|^{\nu}} \frac{I_0(\tau_1 \tau_2)}{f_\infty(-\tau_1^2, -\tau_2^2)}\,. \qquad (3.161)$$

This expression satisfies the normalization condition $\lim_{\tau_2 \to \tau_1} p_s^{(N_f=2)}(\tau_1, \tau_2) = 1$.

In Fig. 3.11 we plot $p_s^{(N_f=2)}(\tau_1, \tau_2)$ as a function of $\tau_2$. The results indicate that larger topology mitigates the sign problem but becomes less effective for heavier masses. This phenomenon is easily understood if one recalls that a large $\nu$ that causes a depletion of eigenvalues near the origin can smoothen the strongly oscillating domain of the spectrum if the domain is close to the origin; since the domain moves away from the origin for heavier masses, the smoothing by $\nu$ becomes less effective.

As a result, the average sign turns out to be essentially independent of topology for larger masses, where the sign problem becomes severe for $|\tau_1 - \tau_2| \gtrsim 2.5$, as can also be seen from Fig. 3.11. In physical units we have, using (3.78),

$$|\delta m_{\text{phys}}| \gtrsim \frac{4.5}{\sqrt{V_4 \Delta^2}} . \tag{3.162}$$

## References


1. N. Yamamoto, T. Kanazawa, Dense QCD in a finite volume. Phys. Rev. Lett. **103**, 032001 (2009)
2. H. Leutwyler, A.V. Smilga, Spectrum of Dirac operator and role of winding number in QCD. Phys. Rev. **D46**, 5607–5632 (1992)
3. R. Casalbuoni, R. Gatto, Effective theory for color-flavor locking in high density QCD. Phys. Lett. **B464**, 111–116 (1999)
4. D.T. Son, M.A. Stephanov, Inverse meson mass ordering in color-flavor-locking phase of high density QCD. Phys. Rev. **D61**, 074012 (2000)
5. D.T. Son, M.A. Stephanov, Inverse meson mass ordering in color-flavor-locking phase of high density QCD: Erratum. Phys. Rev. **D62**, 059902 (2000)
6. T. Schäfer, Mass terms in effective theories of high density quark matter. Phys. Rev. **D65**, 074006 (2002)
7. P.F. Bedaque, T. Schäfer, High density Quark matter under stress. Nucl. Phys. **A697**, 802–822 (2002)
8. P.F. Bedaque, T. Schafer, High density Quark matter under stress. Nucl. Phys. **A697**, 802–822 (2002)
9. T. Schafer, Instanton effects in QCD at high baryon density. Phys. Rev. **D65**, 094033 (2002)
10. N. Yamamoto, Instanton-induced crossover in dense QCD. JHEP **12**, 060 (2008)
11. T. Schäfer, F. Wilczek, Continuity of quark and hadron matter. Phys. Rev. Lett. **82**, 3956–3959 (1999)
12. A.D. Jackson, M.K. Sener, J.J.M. Verbaarschot, Finite volume partition functions and Itzykson-Zuber integrals. Phys. Lett. **B387**, 355–360 (1996)
13. E.V. Shuryak, J.J.M. Verbaarschot, Random matrix theory and spectral sum rules for the Dirac operator in QCD. Nucl. Phys. **A560**, 306–320 (1993)
14. J.J.M. Verbaarschot, I. Zahed, Spectral density of the QCD Dirac operator near zero virtuality. Phys. Rev. Lett. **70**, 3852–3855 (1993)
15. J.C. Osborn, D. Toublan, J.J.M. Verbaarschot, From chiral random matrix theory to chiral perturbation theory. Nucl. Phys. **B540**, 317–344 (1999)
16. P.H. Damgaard, J.C. Osborn, D. Toublan, J.J.M. Verbaarschot, The microscopic spectral density of the QCD Dirac operator. Nucl. Phys. **B547**, 305–328 (1999)
17. Y. Watanabe, K. Fukushima, T. Hatsuda, Order parameters with higher dimensionful composite fields. Prog. Theoret. Phys. **111**, 967–972 (2004)
18. T. Kanazawa, T. Wettig, N. Yamamoto, Chiral Lagrangian and spectral sum rules for dense two-color QCD. JHEP **08**, 003 (2009)
19. T. Kanazawa, T. Wettig, N. Yamamoto, Chiral Lagrangian and spectral sum rules for two-color QCD at high density. PoS LAT2009, 195 (2009)
20. L. He, Nambu-Jona-Lasinio model description of weakly interacting Bose condensate and BEC-BCS crossover in dense QCD-like theories. Phys. Rev. **D82**, 096003 (2010)
21. K. Fukushima, K. Iida, Larkin-Ovchinnikov-Fulde-Ferrell state in two-color quark matter. Phys. Rev. **D76**, 054004 (2007)

22. J.O. Andersen, T. Brauner, Phase diagram of two-color quark matter at nonzero baryon and isospin density. Phys. Rev. **D81**, 096004 (2010)
23. T. Zhang, T. Brauner, D.H. Rischke, QCD-like theories at nonzero temperature and density. JHEP **06**, 064 (2010)
24. T. Brauner, K. Fukushima, Y. Hidaka, Two-color quark matter: $U(1)_A$ restoration, superfluidity, and quarkyonic phase. Phys. Rev. **D80**, 074035 (2009)
25. K. Splittorff, D.T. Son, M.A. Stephanov, QCD-like theories at finite Baryon and Isospin density. Phys. Rev. **D64**, 016003 (2001)
26. D.M. Eagles, Possible pairing without superconductivity at low carrier concentrations in bulk and thin-film superconducting semiconductors. Phys. Rev. **186**, 456–463 (1969)
27. A.J. Leggett, Cooper pairing in spin-polarized Fermi systems, J. Phys. (Paris) Colloq. 41, 7–19 (1980)
28. P. Nozieres, S. Schmitt-Rink, Bose condensation in an attractive fermion gas: From weak to strong coupling superconduct ivity. J. Low. Temp. Phys. **59**, 195 (1985)
29. S. Hands, S. Kim, J.-I. Skullerud, Deconfinement in dense 2-color QCD. Eur. Phys. J. **C48**, 193 (2006)
30. S. Hands, S. Kim, J.-I. Skullerud, A Quarkyonic phase in dense two color matter? Phys. Rev. **D81**, 091502 (2010)
31. M.G. Alford, K. Rajagopal, F. Wilczek, QCD at finite baryon density: nucleon droplets and color superconductivity. Phys. Lett. **B422**, 247–256 (1998)
32. R. Rapp, T. Schäfer, E.V. Shuryak, M. Velkovsky, Diquark Bose condensates in high density matter and instantons. Phys. Rev. Lett. **81**, 53–56 (1998)
33. N. Yamamoto, M. Tachibana, T. Hatsuda, G. Baym, Phase structure, collective modes, and the axial anomaly in dense QCD. Phys. Rev. **D76**, 074001 (2007)
34. T. Schäfer, QCD and the eta' mass: instantons or confinement? Phys. Rev. **D67**, 074502 (2003)
35. D.H. Rischke, Debye screening and Meissner effect in a two-flavor color superconductor. Phys. Rev. **D62**, 034007 (2000)
36. D.H. Rischke, D.T. Son, M.A. Stephanov, Asymptotic deconfinement in high-density QCD. Phys. Rev. Lett. **87**, 062001 (2001)
37. J. Gasser, H. Leutwyler, Thermodynamics of chiral symmetry. Phys. Lett. **B188**, 477 (1987)
38. A.V. Smilga, J.J.M. Verbaarschot, Spectral sum rules and finite volume partition function in gauge theories with real and pseudoreal fermions. Phys. Rev. **D51**, 829–837 (1995)
39. M. Kieburg, T. Guhr, A new approach to derive Pfaffian structures for random matrix ensembles. J. Phys. **A43**, 135204 (2010)
40. J.J.M. Verbaarschot, T. Wettig, Random matrix theory and chiral symmetry in QCD. Ann. Rev. Nucl. Part. Sci. **50**, 343–410 (2000)
41. P.H. Damgaard, S.M. Nishigaki, Universal spectral correlators and massive Dirac operators. Nucl. Phys. **B518**, 495–512 (1998)
42. P.H. Damgaard, Massive spectral sum rules for the Dirac operator. Phys. Lett. **B425**, 151–157 (1998)
43. T. Wilke, T. Guhr, T. Wettig, The microscopic spectrum of the QCD Dirac operator with finite quark masses. Phys. Rev. **D57**, 6486–6495 (1998)
44. T. Kanazawa, T. Wettig, N. Yamamoto, Chiral random matrix theory for two-color QCD at high density. Phys. Rev. **D81**, 081701 (2010)
45. U. Magnea, Random matrices beyond the Cartan classification. J. Phys. **A41**, 045203 (2008)
46. T. Schäfer, Instanton effects in QCD at high baryon density. Phys. Rev. **D65**, 094033 (2002)
47. J.J.M. Verbaarschot, The Spectrum of the Dirac operator near zero virtuality for N(c) = 2 and chiral random matrix theory. Nucl. Phys. **B426**, 559–574 (1994)
48. A.M. Halasz, J.J.M. Verbaarschot, Effective Lagrangians and chiral random matrix theory. Phys. Rev. **D52**, 2563–2573 (1995)
49. B. Vanderheyden, A.D. Jackson, A random matrix model for color superconductivity at zero chemical potential. Phys. Rev. **D61**, 076004 (2000)
50. B. Vanderheyden, A.D. Jackson, Random matrix model for chiral symmetry breaking and color superconductivity in QCD at finite density. Phys. Rev. **D62**, 094010 (2000)

51. B. Vanderheyden, A.D. Jackson, Random matrix study of the phase structure of QCD with two colors. Phys. Rev. **D64**, 074016 (2001)
52. S. Pepin, A. Schäfer, QCD at high baryon density in a random matrix model. Eur. Phys. J. **A10**, 303–308 (2001)
53. B. Klein, D. Toublan, J.J.M. Verbaarschot, The QCD phase diagram at nonzero temperature, baryon and isospin chemical potentials in random matrix theory. Phys. Rev. **D68**, 014009 (2003)
54. B. Klein, D. Toublan, J.J.M. Verbaarschot, Diquark and pion condensation in random matrix models for two-color QCD. Phys. Rev. **D72**, 015007 (2005)
55. J.B. Kogut, M.A. Stephanov, D. Toublan, J.J.M. Verbaarschot, A. Zhitnitsky, QCD-like theories at finite baryon density. Nucl. Phys. **B582**, 477–513 (2000)
56. D.T. Son, Superconductivity by long-range color magnetic interaction in high-density quark matter. Phys. Rev. **D59**, 094019 (1999)
57. G. Akemann, M.J. Phillips, H.J. Sommers, Characteristic polynomials in real ginibre ensembles. J. Phys. **A42**, 012001 (2009)
58. G. Akemann, M.J. Phillips, H.J. Sommers, The chiral Gaussian two-matrix ensemble of real asymmetric matrices. J. Phys. **A43**, 085211 (2010)
59. T. Kanazawa, T. Wettig, N. Yamamoto, Singular values of the Dirac operator in dense QCD-like theories. JHEP **12**, 007 (2011)
60. H.J. Sommers, Symplectic structure of the real Ginibre ensemble. J. Phys. **A40**, F671 (2007)
61. P.J. Forrester, T. Nagao, Eigenvalue statistics of the real Ginibre ensemble. Phys. Rev. Lett. **99**, 050603 (2007)
62. A. Borodin, C.D. Sinclair, Correlation functions of asymmetric real matrices, http://xxx.lanl.gov/abs/0706.2670 arXiv:0706.2670
63. A. Borodin, C.D. Sinclair, The Ginibre ensemble of real random matrices and its scaling limits. Commun. Math. Phys. **291**, 177–224 (2009)
64. P.J. Forrester, T. Nagao, Skew orthogonal polynomials and the partly symmetric real Ginibre ensemble. J. Phys. **A41**, 375003 (2008)
65. H.J. Sommers, W. Wieczorek, General eigenvalue correlations for the real ginibre ensemble. J. Phys. **A41**, 405003 (2008)
66. G. Akemann, M. Kieburg, M.J. Phillips, Skew-orthogonal Laguerre polynomials for chiral real asymmetric random matrices. J. Phys. **A43**, 375207 (2010)
67. M. Kieburg, T. Guhr, Derivation of determinantal structures for random matrix ensembles in a new way. J. Phys. **A43**, 075201 (2010)
68. G. Akemann, Matrix models and QCD with chemical potential. Int. J. Mod. Phys. **A22**, 1077–1122 (2007)
69. T. Nagao, S.M. Nishigaki, Massive chiral random matrix ensembles at beta = 1 and 4: finite-volume QCD partition functions. Phys. Rev. **D62**, 065006 (2000)
70. T. Nagao, S.M. Nishigaki, Massive chiral random matrix ensembles at beta = 1 and 4: QCD Dirac operator spectra. Phys. Rev. **D62**, 065007 (2000)
71. J.C. Osborn, K. Splittorff, J.J.M. Verbaarschot, Chiral symmetry breaking and the Dirac spectrum at nonzero chemical potential. Phys. Rev. Lett. **94**, 202001 (2005)
72. J.C. Osborn, K. Splittorff, J.J.M. Verbaarschot, Chiral condensate at nonzero chemical potential in the microscopic limit of QCD. Phys. Rev. **D78**, 065029 (2008)
73. J.C. Osborn, K. Splittorff, J.J.M. Verbaarschot, Phase diagram of the Dirac spectrum at nonzero chemical potential. Phys. Rev. **D78**, 105006 (2008)
74. G. Akemann, T. Kanazawa, M.J. Phillips, T. Wettig, Random matrix theory of unquenched two-colour QCD with nonzero chemical potential. JHEP **03**, 066 (2011)
75. M.J. Phillips, A random matrix model for two-colour QCD at non-zero quark density. Doctoral thesis submitted to Brunel University (2011) http://bura.brunel.ac.uk/handle/2438/5084
76. J.C.R. Bloch, T. Wettig, Random matrix analysis of the QCD sign problem for general topology. JHEP **03**, 100 (2009)
77. K. Splittorff, J.J.M. Verbaarschot, Phase of the Fermion determinant at nonzero chemical potential. Phys. Rev. Lett. **98**, 031601 (2007)

78. K. Splittorff, J.J.M. Verbaarschot, The QCD sign problem for small chemical potential. Phys. Rev. **D75**, 116003 (2007)
79. K. Splittorff, The sign problem in the epsilon-regime of QCD. PoS LAT2006, 023 (2006).
80. J. Han, M.A. Stephanov, A random matrix study of the QCD sign problem. Phys. Rev. **D78**, 054507 (2008)
81. G. Akemann, G. Vernizzi, Characteristic polynomials of complex random matrix models. Nucl. Phys. **B660**, 532–556 (2003)
82. S. Hands et al., Numerical study of dense adjoint matter in two color QCD. Eur. Phys. J. **C17**, 285–302 (2000)
83. S. Hands, I. Montvay, L. Scorzato, J. Skullerud, Diquark condensation in dense adjoint matter. Eur. Phys. J. **C22**, 451–461 (2001)

# Chapter 4
# Three-fold Way at High Density

## 4.1 Introduction

As we reviewed in Chap. 2, the patterns of chiral symmetry breaking in QCD at zero density can be classified into three types (cf. Table 2.2) according to the color representation of fermions. As noticed by Verbaarschot et al., there exist three symmetry classes in the chiral gaussian ensemble of random matrices, in one-to-one correspondence with the three types of symmetry breaking on the QCD side. They are called ChGOE, ChGUE and ChGSE, associated with the Dyson index $\beta = 1$, 2 and 4, respectively. This correspondence between Chiral random matrix theory (ChRMT) and QCD is called the "three-fold way" after the well-known three classes in the Wigner-Dyson ensemble. Although the inclusion of nonzero chemical potential ($\mu$) into this scheme has been attempted so far (see [1] for a review), the existing literature is entirely limited to the $\varepsilon$-regime at low density where $\mu$ must go to zero in the thermodynamic limit in such a way that $V_4\ F_\pi^2 \mu^2 \sim \mathcal{O}(1)$.

In Chap. 3, we elucidated how to overcome this limitation: there exists a new non-Hermitian ChRMT corresponding to the high-density BCS phase of two-color QCD (with an even number of flavors). Moreover it can be shown in a straightforward manner that the same ChRMT applies to all dense QCD-like theories with fermions in a *pseudoreal* representation of the gauge group ($\beta = 1$).[1,2]

Our goal in this chapter is to generalize our scheme to the other two classes of QCD-like theories at high density: QCD with complex fermions at high isospin density ($\beta = 2$) and QCD with real quarks at high quark number density ($\beta = 4$). The patterns of spontaneous symmetry breaking and associated ChRMTs are summarized in Tables 4.1 and 4.2, respectively. We will show the correspondence of ChRMTs to dense QCD-like theories through bosonization, as we did in Sect. 3.3. By doing this

[1] See footnote 2 on p. 14 for the mathematical definition of real and pseudoreal representations.

[2] For a generic pseudoreal representation we have to replace $\varepsilon_{ab}$ within the diquark pairing (3.19) by a proper antisymmetric matrix of a given gauge group. However it does not affect the symmetry breaking pattern (3.20) and the correspondence to the ChRMT (3.74).

T. Kanazawa, *Dirac Spectra in Dense QCD*, Springer Theses,
DOI: 10.1007/978-4-431-54165-3_4, © Springer Japan 2013

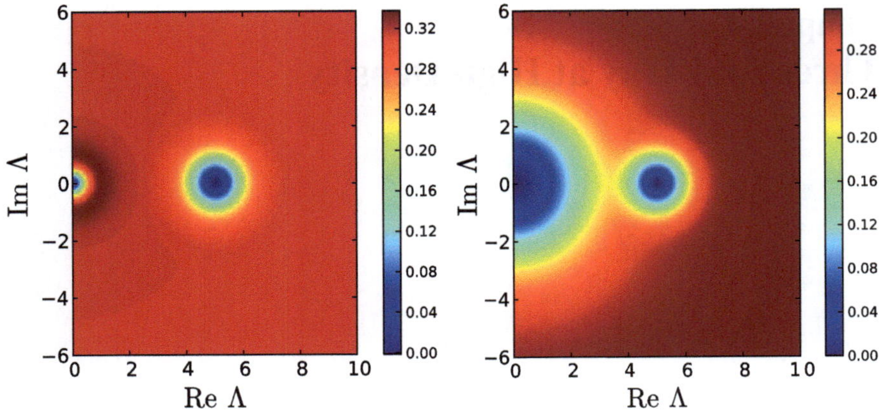


**Fig. 4.1** The colormap plot of $\rho_{2,\nu}(\Lambda)$ for $\tau = \check{\tau} = 5$ with $\nu = 0$ (*left*) and $\nu = 10$ (*right*)

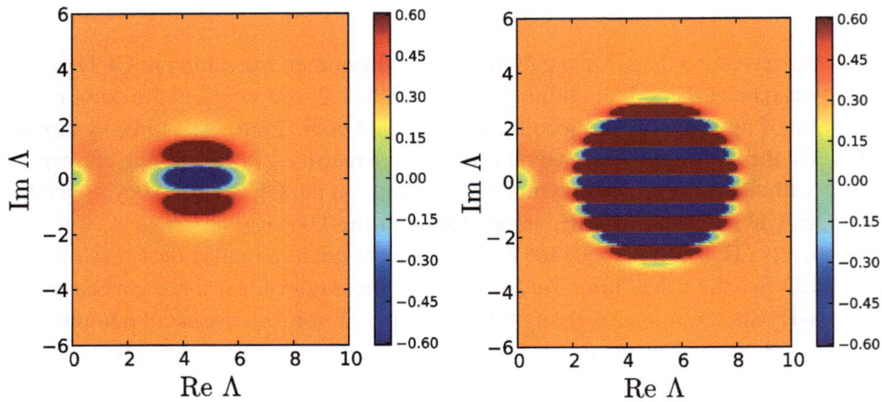


**Fig. 4.2** The colormap plot of $\rho_{2,\nu}(\Lambda)$ for $(\tau, \check{\tau}) = (3, 6)$ (*left*) and $(\tau, \check{\tau}) = (2, 8)$ (*right*) with $\nu = 0$. (The values outside the range $[-0.6, 0.6]$ were clipped for a better visibility of the figure.)

we extend the "three-fold way" of matrix models at low density to the BCS superfluid phase at high density. For $\beta = 2$ we also work out the microscopic spectral density in the large-$N$ limit exactly and find behaviors similar to two-color QCD ($\beta = 1$) as the masses were varied.

## 4.2 QCD at Large Isospin Density ($\beta = 2$)

### *4.2.1 QCD Side*

The theory we consider as a canonical example is SU($N_c$) gauge theory with $N_c \geq 3$ colors and $2N_f$ Dirac fermions in the fundamental representation of SU($N_c$), with chemical potential $-\mu$ for $N_f$ flavors ($=u^f$) and $+\mu$ for the other $N_f$ flavors ($=d^f$).

**Table 4.1** The 'three-fold way' of spontaneous symmetry breaking in QCD at high density with $N_f$ Dirac fermions

| Color rep. of quarks | Symmetry Breaking Pattern |
|---|---|
| Pseudoreal | $\mathrm{U}(N_f)_R \times \mathrm{U}(N_f)_L \to \mathrm{Sp}(N_f)_R \times \mathrm{Sp}(N_f)_L$ |
| Complex | $\mathrm{U}(N_f)_{u_R} \times \mathrm{U}(N_f)_{u_L} \times \mathrm{U}(N_f)_{d_R} \times \mathrm{U}(N_f)_{d_L}$ $\to \mathrm{U}(N_f)_{u_R+d_L} \times \mathrm{U}(N_f)_{u_L+d_R}$ |
| Real | $\mathrm{U}(N_f)_R \times \mathrm{U}(N_f)_L \to \mathrm{SO}(N_f)_R \times \mathrm{SO}(N_f)_L$ |

For complex fermions (the real/pseudoreal fermions), the chemical potential for the isospin density (for the quark number density) is considered. For pseudoreal fermions, $N_f$ is assumed to be even. The $\mathrm{U}(1)_A$ symmetry is restored at high density due to medium effects. Compare with the three-fold way at zero density in Table 2.2

**Table 4.2** The 'three-fold way' of ChRMT for QCD at high density

| Color rep. of quarks | Dirac operator in RMT | Matrix elements of $X$ and $Y$ |
|---|---|---|
| Pseudoreal | $\begin{pmatrix} 0 & X \\ Y & 0 \end{pmatrix}$ | real ($\beta = 1$) |
| Complex | | complex ($\beta = 2$) |
| Real | | real quaternion ($\beta = 4$) |

The blocks $X$ and $Y$ are independent gaussian random matrices. See the caption of Table 4.1 for more details

This includes three-color QCD with $2N_f = 2$ at finite isospin chemical potential as a special case. This theory, also known as the *phase-quenched QCD* [2], is free from the sign problem for equal masses [3] and a direct lattice simulation is possible [4], providing an opportunity to test theoretical predictions based on ChPT [5–10]. The usefulness of ChPT in probing the phase diagram originates from the fact that, as in two-color QCD, the global symmetry of this theory is explicitly broken by nonzero chemical potential.

### Ground State and Symmetry Breaking at $\mu \gg \Lambda$

Let us denote the first $N_f$ flavors by $u^f$ ($f = 1, \ldots, N_f$) and the second $N_f$ flavors by $d^f$ ($f = 1, \ldots, N_f$). First of all, the global symmetry of the Lagrangian at $\mu = 0$ in the chiral limit is $\mathrm{U}(2N_f)_R \times \mathrm{U}(2N_f)_L$. The nonzero $\mu$ breaks this symmetry explicitly down to

$$G = \mathrm{U}(N_f)_{u_R} \times \mathrm{U}(N_f)_{u_L} \times \mathrm{U}(N_f)_{d_R} \times \mathrm{U}(N_f)_{d_L}, \tag{4.1}$$

with an obvious notation. At sufficiently high density and for small quark masses $m_f \ll \Delta$, the Fermi surfaces of $\bar{u}$ and $d$ are degenerate and the attractive interaction between them will lead to the formation of the bilinear condensate, owing to the BCS mechanism [5, 6, 11, 12]:

$$\langle \bar{u}^f \gamma_5 d^g \rangle \sim \frac{\mu^2 \Delta}{g} \delta^{fg} \quad \text{with} \quad \Delta \sim \frac{\mu}{g^5} \exp\left( -\frac{\pi^2}{g} \sqrt{\frac{6N_c}{N_c^2 - 1}} \right), \tag{4.2}$$

which breaks parity. While $\bar{u}\gamma_5 d(= 0^-)$ and $\bar{u}d(= 0^+)$ are not distinguished in the one-gluon exchange interaction, the non-perturbative instanton-induced interaction and nonzero quark masses favor $\bar{u}\gamma_5 d$ over $\bar{u}d$ and the degeneracy gets lifted [13, 14]. We note that, for $2N_f = 2$, (4.2) has the same quantum numbers as the pion condensate at low density. This observation hints at a continuous crossover from low to high density [5, 6]. Since (4.2) is a color singlet, the ground state will be a BCS superfluid in the confining phase, similarly to two-color QCD at high density; the color superconductivity does not take place.

The condensate (4.2) breaks the symmetry $G$ down to

$$H = \mathrm{U}(N_f)_{u_R + d_L} \times \mathrm{U}(N_f)_{u_L + d_R} \,. \tag{4.3}$$

The spontaneous symmetry breaking $G \to H$ gives rise to $2N_f^2$ NG modes with the quantum numbers

$$\bar{u}\gamma_5 \tau^A d, \quad \bar{d}\gamma_5 \tau^A u \tag{4.4}$$

where $\{\tau^A\}$ are the generators of $\mathrm{U}(N_f)$. The chiral anomaly can be ignored at sufficiently large $\mu$ due to asymptotic freedom and the screening of instantons by dense medium.

Since the measure of this theory is positive definite for equal masses, i.e.,

$$\det{}^{N_f} (\mathcal{D}(\mu) + m) \det{}^{N_f} (\mathcal{D}(-\mu) + m) = |\det (\mathcal{D}(\mu) + m)|^{2N_f} > 0 \,, \tag{4.5}$$

the method of QCD inequalities [15, 16] can be applied. Generalizing the argument for the $2N_f = 2$ case in [5, 6] to generic $N_f$ is straightforward and one can verify that the symmetry breaking driven by the condensate (4.2) is fully consistent with the constraints from QCD inequalities.

For unequal masses, the Fermi surfaces of different flavors are not coincident and the pairing (4.2) is not necessarily the most favored one. If the discrepancy of masses is as large as $\sqrt{\mu\Delta}$, inhomogeneous phases will be realized [17]. In the present work we avoid such exotic phases by always assuming ${}^\forall m_f \ll \Delta$.

### Chiral Lagrangian at High Density

The low-energy effective theory can be constructed via the so-called spurion analysis, as in Sect. 3.2.1. We introduce coset fields which are related to the quark fields as

$$U_1^{fg} \sim d_L^f \bar{u}_R^g, \quad U_2^{fg} \sim u_L^f \bar{d}_R^g \,. \tag{4.6}$$

These fields live on the coset space $G/H \cong \mathrm{U}(N_f) \times \mathrm{U}(N_f)$. Under a global unitary transformation

$$u_R \to g_{u_R} u_R \tag{4.7}$$
$$u_L \to g_{u_L} u_L \tag{4.8}$$
$$d_R \to g_{d_R} d_R \tag{4.9}$$
$$d_L \to g_{d_L} d_L \tag{4.10}$$

the coset elements transform as

$$U_1 \to g_{d_L} U_1 g_{u_R}^\dagger, \quad U_2 \to g_{u_L} U_2 g_{d_R}^\dagger \tag{4.11}$$

while the mass matrix must transform as

$$M_1 \to g_{u_L} M_1 g_{u_R}^\dagger, \quad M_2 \to g_{d_L} M_2 g_{d_R}^\dagger \tag{4.12}$$

to keep the mass term in the Lagrangian $\bar{u}_L M_1 u_R + \bar{u}_R M_1^\dagger u_L + \bar{d}_L M_2 d_R + \bar{d}_R M_2^\dagger d_L$ invariant.

Therefore the static part of the effective Lagrangian at $\mathcal{O}(M^2)$ is uniquely determined to be

$$\mathcal{L}_{\text{stat}} = -\Lambda^2 \mathrm{tr}[M_1 U_1^\dagger M_2 U_2^\dagger] + \text{c.c.} \tag{4.13}$$

where $\Lambda$ is a low energy constant of mass dimension 1. The sign was chosen so that $U_1 = U_2 = \mathbf{1}$ is the ground state. There is no invariant term at $\mathcal{O}(M)$.

The constant terms $\mathrm{tr}[M_1^\dagger M_1]$ and $\mathrm{tr}[M_2^\dagger M_2]$ are neglected as they do not affect the dynamics of the NG modes (see also the Appendix A.3). The effective chemical potential of $\mathcal{O}(M^2)$ (the so-called Bedaque-Schäfer term in the CFL phase of three-color QCD [18]) is also dropped since it is suppressed by $1/\mu$ at large $\mu$.

## Matching with HDET

We now show that the low energy constant $\Lambda$ in $\mathcal{L}_{\text{stat}}$ is equal to $\Delta$ (the BCS gap) up to a numerical constant through the matching with the high density effective theory (HDET) [19, 20]. Although the procedure is essentially the same as in the two-color case (Appendix A.2), there is a subtlety in the present case that calls for a separate treatment. As discussed in the Appendix A.2 the mass term induces the following four-fermion vertex in HDET [21]:

$$\begin{aligned} \mathcal{L}_{\text{mass}} = &\frac{g^2}{8\mu^4} (\psi_{i,L}^{\dagger\, a} C \psi_{j,L}^{*\, b})(\psi_{k,R}^{T\, c} C \psi_{l,R}^{d}) \\ &\times \left[ (T^A)_{ac} (T^A)_{bd} M_{ik} M_{jl} \right] + (L \leftrightarrow R,\ M \leftrightarrow M^\dagger), \end{aligned} \tag{4.14}$$

where $i, j, k, l$ $(a, b, c, d)$ are flavor (color) indices. Since this is derived for fermions with equal chemical potential $\mu$ we cannot blindly apply it to the case of isospin chemical potential. As a trick let us introduce a conjugate field of $u^f$, say $u_c^f$,

$$u_c^f = C(\bar{u}^f)^T \quad (f = 1, 2, \ldots, N_f) \tag{4.15}$$

with which the Lagrangian is cast into the form

$$\mathcal{L} = \bar{u}[\mathcal{D}(-\mu) + M_1 P_R + M_1^\dagger P_L]u + \bar{d}[\mathcal{D}(\mu) + M_2 P_R + M_2^\dagger P_L]d \tag{4.16}$$

$$= \bar{u}_c[\mathcal{D}_{\mathrm{AF}}(\mu) + \tilde{M}_1 P_R + \tilde{M}_1^\dagger P_L]u_c + \bar{d}[\mathcal{D}(\mu) + M_2 P_R + M_2^\dagger P_L]d\,. \tag{4.17}$$

Here $\tilde{M}_1 \equiv M_1^T$ is the mass matrix for $u_c$ whereas $\mathcal{D}_{\mathrm{AF}}$ denotes the Dirac operator in the *anti-fundamental* representation of the gauge group. The virtue of this rewriting is that in the new basis, all quarks have a common chemical potential $\mu$, so that we can now use (4.14) to compute the shift of the ground state energy density $\delta\mathcal{E}$ due to the mass term. Namely we obtain

$$\delta\mathcal{E} \propto \frac{g^2}{\mu^4}\Big\langle \big((u_c^\dagger)^a_{i,L} C d^{*b}_{j,L}\big)\big((u_c^T)^c_{k,R} C d^d_{l,R}\big)(\tilde{T}^A)_{ac}(T^A)_{bd}(\tilde{M}_1)_{ik}(M_2)_{jl} + (L \leftrightarrow R,\ M \leftrightarrow M^\dagger)\Big\rangle \tag{4.18}$$

with $\tilde{T}^A \equiv -(T^A)^*$ the color generator in the *anti-fundamental* representation. Using $\langle (u_c)^a_f C d^b_g \rangle \propto \delta_{fg}\delta^{ab}$ as well as (4.2) we find

$$\delta\mathcal{E} \propto \frac{g^2}{\mu^4}\left(\frac{\mu^2\Delta}{g}\right)^2 \delta_{ij}\delta^{ab}\delta_{kl}\delta^{cd}(\tilde{T}^A)_{ac}(T^A)_{bd}(\tilde{M}_1)_{ik}(M_2)_{jl} + \text{c.c.} \tag{4.19}$$

$$= \Delta^2 \mathrm{tr}[\tilde{M}_1^T M_2]\mathrm{tr}[(\tilde{T}^A)^T T^A] + \text{c.c.} \tag{4.20}$$

$$\propto -\Delta^2 \mathrm{tr}[M_1 M_2] + \text{c.c.} \tag{4.21}$$

On the other hand, the chiral effective theory (4.13) evaluated at the ground state ($U_1 = U_2 = \mathbf{1}$) gives $\delta\mathcal{E} = -\Lambda^2\mathrm{tr}[M_1 M_2] + \text{c.c.}$ The comparison shows that $\Lambda$ is equal to $\Delta$ up to a numerical factor, as claimed.

### The $\varepsilon$-Regime

The $\varepsilon$-regime (or the microscopic domain) of this theory can be defined just as in two-color QCD. The masses of the NG modes $m_{\mathrm{NG}}$ are of order $m_f\Delta/\mu$ (cf. Eq. 3.36), whereas the mass of the lightest non-NG particle is $\sim \Delta$. If the system is put in a four-dimensional Euclidean box of size $L \times L \times L \times \beta \sim L^4$ with $\beta = 1/T$ satisfying

$$\frac{1}{\Delta} \ll L \ll \frac{1}{m_{\rm NG}}, \tag{4.22}$$

then the zero-momentum modes of the NG modes will dominate the partition function. In this domain, the kinetic term of the effective theory drops out and the finite-volume partition function is simply given by

$$Z_{2N_f}^{\rm QCD}(M_1, M_2) = \int_{\mathrm{U}(N_f)} [dU_1] \int_{\mathrm{U}(N_f)} [dU_2] \exp\left(\# V_4 \Delta^2 \mathrm{Re}\,\mathrm{tr}[M_1 U_1^\dagger M_2 U_2^\dagger]\right), \tag{4.23}$$

where we are cavalier about the $\mathcal{O}(1)$ factor in the exponent. Since the $\mathrm{U}(1)_A$ anomaly is ignored, the partition function is independent of the $\theta$ angle (the replacement $M_i \to M_i\,\mathrm{e}^{i\theta/N_f}$ is absorbed in the invariance of the Haar measure) and the topologically non-trivial sectors of the gauge fields do not contribute to (4.23). The extension to the intermediate density region where $\mathrm{U}(1)_A$ anomaly is not negligible is discussed in [22].

### 4.2.2 RMT Side

The ChRMT we consider is defined by

$$\begin{aligned} Z_{2N_f,\nu}^{\rm RMT}(\{m_f\}, \{\check{m}\}_f\}) = & \int_{\mathbb{C}^{N(N+\nu)}} [dP] \int_{\mathbb{C}^{N(N+\nu)}} [dQ]\; \mathrm{e}^{-N\mathrm{tr}[PP^\dagger + QQ^\dagger]} \\ & \times \prod_{f=1}^{N_f} \det\begin{pmatrix} m_f^* \mathbf{1}_N & P \\ Q^\dagger & m_f \mathbf{1}_{N+\nu} \end{pmatrix} \prod_{g=1}^{N_f} \det\begin{pmatrix} \check{m}_g^* \mathbf{1}_N & Q \\ P^\dagger & \check{m}_g \mathbf{1}_{N+\nu} \end{pmatrix}, \end{aligned} \tag{4.24}$$

where $P$ and $Q$ are complex $N \times (N+\nu)$ matrices ($\beta = 2$). The integer $\nu$ is arbitrary.

In the chiral limit, the determinants can be rearranged as

$$\det{}^{N_f}\begin{pmatrix} \mathbf{0} & P \\ -P^\dagger & \mathbf{0} \end{pmatrix} \det{}^{N_f}\begin{pmatrix} \mathbf{0} & Q \\ -Q^\dagger & \mathbf{0} \end{pmatrix}, \tag{4.25}$$

which suggests (cf. Sect. 2.3.3) that the spontaneous symmetry breaking of this model at large $N$ must take place in such a way that

$$[\mathrm{U}(N_f) \times \mathrm{U}(N_f)]_P \times [\mathrm{U}(N_f) \times \mathrm{U}(N_f)]_Q \to [\mathrm{U}(N_f)_V]_P \times [\mathrm{U}(N_f)_V]_Q. \tag{4.26}$$

To our delight, this pattern exactly coincides with the symmetry breaking $G \to H$ on the QCD side (see (4.1) and (4.3)); we are moving in the right direction!

### The $\sigma$-Model Representation: Matching with QCD

We define the $N_f \times N_f$ mass matrices $M_1$, $M_2$ by

$$M_1 \equiv \mathrm{diag}(m_1, \dots, m_{N_f}), \quad M_2 \equiv \mathrm{diag}(\check{m}_1, \dots, \check{m}_{N_f}). \tag{4.27}$$

With the help of the fermionic variables $u, \bar{u}, d, \bar{d}$, the partition function can be expressed as

$$\begin{aligned} Z^{\mathrm{RMT}}_{2N_f,\nu}(M_1, M_2) &= \iint [dP][dQ]\, \mathrm{e}^{-N\mathrm{tr}[PP^\dagger + QQ^\dagger]} \int d[u, \bar{u}, d, \bar{d}] \\ &\quad \times \exp\left[ \begin{pmatrix} \bar{u}_R \\ \bar{u}_L \end{pmatrix} \begin{pmatrix} M_1^\dagger \otimes \mathbf{1}_N & \mathbf{1}_{N_f} \otimes P \\ \mathbf{1}_{N_f} \otimes Q^\dagger & M_1 \otimes \mathbf{1}_{N+\nu} \end{pmatrix} \begin{pmatrix} u_L \\ u_R \end{pmatrix} \right. \\ &\quad \left. + \begin{pmatrix} \bar{d}_R \\ \bar{d}_L \end{pmatrix} \begin{pmatrix} M_2^\dagger \otimes \mathbf{1}_N & \mathbf{1}_{N_f} \otimes Q \\ \mathbf{1}_{N_f} \otimes P^\dagger & M_2 \otimes \mathbf{1}_{N+\nu} \end{pmatrix} \begin{pmatrix} d_L \\ d_R \end{pmatrix} \right]. \end{aligned} \tag{4.28}$$

The exponent may be decomposed into three pieces: ($P$ part) + ($Q$ part) + (mass terms). The '$P$ part' reads

$$\begin{aligned} & - N P_{ij} P^*_{ij} + \bar{u}^f_{Ri} P_{ij} u^f_{Rj} + \bar{d}^f_{Lj} P^*_{ij} d^f_{Li} \\ &= -N \left( P_{ij} - \frac{1}{N} \bar{d}^f_{Lj} d^f_{Li} \right) \left( P^*_{ij} - \frac{1}{N} \bar{u}^g_{Ri} u^g_{Rj} \right) - \frac{1}{N} (\bar{d}^f_L u^g_R)(\bar{u}^g_R d^f_L)\,. \end{aligned} \tag{4.29}$$

Similarly the '$Q$ part' reads

$$\begin{aligned} & - N Q_{ij} Q^*_{ij} + \bar{u}^f_{Lj} Q^*_{ij} u^f_{Li} + \bar{d}^f_{Ri} Q_{ij} d^f_{Rj} \\ &= -N \left( Q_{ij} - \frac{1}{N} \bar{u}^f_{Lj} u^f_{Li} \right) \left( Q^*_{ij} - \frac{1}{N} \bar{d}^g_{Ri} d^g_{Rj} \right) - \frac{1}{N} (\bar{u}^f_L d^g_R)(\bar{d}^g_R u^f_L)\,. \end{aligned} \tag{4.30}$$

Now we integrate over $P$ and $Q$ and perform the Hubbard-Stratonovich transformation with the help of $N_f \times N_f$ complex auxiliary fields $\sigma_1$ and $\sigma_2$:

$$\begin{aligned} Z^{\mathrm{RMT}}_{2N_f,\nu}(M_1, M_2) &= \int d[u, \bar{u}, d, \bar{d}] \exp\Big[ -\frac{1}{N} (\bar{d}^f_L u^g_R)(\bar{u}^g_R d^f_L) - \frac{1}{N} (\bar{u}^f_L d^g_R)(\bar{d}^g_R u^f_L) \\ &\quad + \bar{u}^f_R (M_1^\dagger)_{fg} u^g_L + \bar{u}^f_L (M_1)_{fg} u^g_R + \bar{d}^f_R (M_2^\dagger)_{fg} d^g_L + \bar{d}^f_L (M_2)_{fg} d^g_R \Big] \end{aligned} \tag{4.31}$$

$$\begin{aligned} &= \int d[\sigma_1, \sigma_2] \int d\big[u, \bar{u}, d, \bar{d}\big] \exp\Big[ - N\mathrm{tr}[\sigma_1^\dagger \sigma_1 + \sigma_2^\dagger \sigma_2] \\ &\quad + \sigma_1^{fg} (\bar{d}^f_L u^g_R) - \sigma_1^{fg*} (\bar{u}^g_R d^f_L) + \sigma_2^{fg} (\bar{u}^f_L d^g_R) - \sigma_2^{fg*} (\bar{d}^g_R u^f_L) \\ &\quad + \bar{u}^f_R (M_1^\dagger)_{fg} u^g_L + \bar{u}^f_L (M_1)_{fg} u^g_R + \bar{d}^f_R (M_2^\dagger)_{fg} d^g_L + \bar{d}^f_L (M_2)_{fg} d^g_R \Big] \end{aligned} \tag{4.32}$$

$$= \int d[\sigma_1, \sigma_2] \int d\big[u, \bar{u}, d, \bar{d}\big] \exp\Big[ - N\mathrm{tr}[\sigma_1^\dagger \sigma_1 + \sigma_2^\dagger \sigma_2]$$

$$+\begin{pmatrix} \bar{u}_R \\ \bar{u}_L \\ \bar{d}_R \\ \bar{d}_L \end{pmatrix} \begin{pmatrix} \mathbf{0} & M_1^\dagger \otimes \mathbf{1}_N & \mathbf{0} & -\sigma_1^\dagger \otimes \mathbf{1}_N \\ M_1 \otimes \mathbf{1}_{N+\nu} & \mathbf{0} & \sigma_2 \otimes \mathbf{1}_{N+\nu} & \mathbf{0} \\ \mathbf{0} & -\sigma_2^\dagger \otimes \mathbf{1}_N & \mathbf{0} & M_2^\dagger \otimes \mathbf{1}_N \\ \sigma_1 \otimes \mathbf{1}_{N+\nu} & \mathbf{0} & M_2 \otimes \mathbf{1}_{N+\nu} & \mathbf{0} \end{pmatrix} \begin{pmatrix} u_R \\ u_L \\ d_R \\ d_L \end{pmatrix} \Bigg] \tag{4.33}$$

$$= \int d[\sigma_1, \sigma_2]\, \mathrm{e}^{-N \mathrm{tr}[\sigma_1^\dagger \sigma_1 + \sigma_2^\dagger \sigma_2]} \det{}^{N+\nu} \begin{pmatrix} M_1 & \sigma_2 \\ \sigma_1 & M_2 \end{pmatrix} \det{}^N \begin{pmatrix} M_1^\dagger & -\sigma_1^\dagger \\ -\sigma_2^\dagger & M_2^\dagger \end{pmatrix} . \tag{4.34}$$

So far all transformations are exact. Next we take the large-$N$ limit in which $\sqrt{N}\|M_{1,2}\| = \mathcal{O}(1)$. The integral may be evaluated with the saddle point method. Using the singular value decomposition for $\sigma_{1,2}$, one finds that the saddle point manifold is parametrized by unitary group elements:

$$\sigma_1 = U_1 \in \mathrm{U}(N_f), \quad \sigma_2 = U_2 \in \mathrm{U}(N_f). \tag{4.35}$$

Plugging this expression into (4.31) and expanding the determinant to the lowest order we finally arrive at the $\sigma$-model representation of this ChRMT:

$$Z_{2N_f,\nu}^{\mathrm{RMT}}(M_1, M_2) \sim \int_{\mathrm{U}(N_f)} [dU_1] \int_{\mathrm{U}(N_f)} [dU_2]\, \det{}^\nu(U_1 U_2) \exp\left(2N \mathrm{Re}\, \mathrm{tr}[M_1 U_1^\dagger M_2 U_2^\dagger]\right) . \tag{4.36}$$

In particular,

$$Z_{2,\nu}^{\mathrm{RMT}}(m, \check{m}) \sim I_\nu(2N m \check{m}) . \tag{4.37}$$

The random matrix partition function (4.36) at $\nu = 0$ is exactly identical to the QCD partition function (4.23) under the identification

$$\sqrt{V_4 \Delta^2} m_f \Big|_{\mathrm{QCD}} \Longleftrightarrow \sqrt{N} m_f \Big|_{\mathrm{RMT}} \tag{4.38}$$

up to a numerical factor. The same mass dependence of $Z^{\mathrm{QCD}}$ and $Z^{\mathrm{RMT}}$ ensures that all spectral sum rules are identical. This is a clear indication of spectral universality.

### Microscopic Spectral Density

In the $\varepsilon$-regime, the spectral correlations of the Dirac eigenvalues in QCD can be exactly mapped to ChRMT. In what follows, we outline the derivation of the microscopic spectral density of the model (4.24) in four steps. This is a new result and is not known in the literature.

For simplicity we assume $\nu \geq 0$ unless stated otherwise.

• **Step 1**: We denote by $\{\Lambda_k^2\}_{k=1}^N$ the $N$ eigenvalues of $PQ^\dagger$. The $2N+\nu$ eigenvalues of the matrix $\begin{pmatrix} \mathbf{0} & P \\ Q^\dagger & \mathbf{0} \end{pmatrix}$ then consist of $\nu$ exact zero eigenvalues and $2N$ nonzero eigenvalues $\{\Lambda_k, -\Lambda_k\}_{k=1}^N$. Using a mathematical result in [23] we can readily find the so-called eigenvalue representation of this model:

$$Z^{\rm RMT}_{2N_f,\nu}(M_1,M_2) = \Big(\prod_{f=1}^{2N_f} m_f^\nu\Big) \int_{\mathbb{C}^N} \prod_{k=1}^N d^2\Lambda_k \times \Big\{ w_\nu(\Lambda_k) \prod_{f=1}^{N_f} (m_f^2 - \Lambda_k^2)(\check{m}_f^2 - \Lambda_k^{*2}) \Big\} \Big|\Delta_N(\Lambda^2)\Big|^2 \tag{4.39}$$

with

$$w_\nu(\Lambda) \equiv |\Lambda|^{2\nu+2} K_\nu(2N|\Lambda|^2) \tag{4.40}$$

and $\Delta_N$ is the Vandermonde determinant.

• **Step 2**: Next, we wish to express the spectral density

$$R_{2N_f,\nu}(\Lambda) \equiv \Big\langle \sum_{i=1}^N \delta^2(\Lambda - \Lambda_i) \Big\rangle \tag{4.41}$$

as a ratio of partition functions. By definition,

$$R_{2N_f,\nu}(\Lambda) = \frac{1}{Z^{\rm RMT}_{2N_f,\nu}(M_1,M_2)} \Big(\prod_{f=1}^{N_f} m_f \check{m}_f\Big)^\nu \times \int_{\mathbb{C}^N} \prod_{k=1}^N d^2\Lambda_k \Big\{ w_\nu(\Lambda_k) \prod_{f=1}^{N_f} (m_f^2 - \Lambda_k^2)(\check{m}_f^2 - \Lambda_k^{*2}) \Big\} \times \Big|\Delta_N(\Lambda^2)\Big|^2 \sum_{i=1}^N \delta^2(\Lambda - \Lambda_i) \tag{4.42}$$

$$= \frac{N}{Z^{\rm RMT}_{2N_f,\nu}(M_1,M_2)} \Big(\prod_{f=1}^{N_f} m_f \check{m}_f\Big)^\nu \times \int_{\mathbb{C}^N} \prod_{k=1}^N d^2\Lambda_k \Big\{ w_\nu(\Lambda_k) \prod_{f=1}^{N_f} (m_f^2 - \Lambda_k^2)(\check{m}_f^2 - \Lambda_k^{*2}) \Big\} \times \Big|\Delta_N(\Lambda^2)\Big|^2 \delta^2(\Lambda - \Lambda_N) \tag{4.43}$$

$$
\begin{aligned}
&= \frac{N}{Z^{\rm RMT}_{2N_f,\nu}(M_1, M_2)} \Big(\prod_{f=1}^{N_f} m_f \check{m}_f\Big)^\nu w_\nu(\Lambda) \prod_{f=1}^{N_f} (m_f^2 - \Lambda^2)(\check{m}_f^2 - \Lambda^{*2}) \\
&\quad \times \int_{\mathbb{C}^{N-1}} \prod_{k=1}^{N-1} d^2\Lambda_k \Big\{ w_\nu(\Lambda_k)(\Lambda^2 - \Lambda_k^2)(\Lambda^{*2} - \Lambda_k^{*2}) \\
&\quad \times \prod_{f=1}^{N_f} (m_f^2 - \Lambda_k^2)(\check{m}_f^2 - \Lambda_k^{*2}) \Big\} \Big| \Delta_{N-1}(\Lambda^2) \Big|^2 \qquad (4.44) \\
&= N \frac{1}{|\Lambda|^{2\nu}} w_\nu(\Lambda) \prod_{f=1}^{N_f} (m_f^2 - \Lambda^2)(\check{m}_f^2 - \Lambda^{*2}) \\
&\quad \times \frac{Z^{\rm RMT}_{2(N_f+1),\nu}(\{M_1, \Lambda\}, \{M_2, \Lambda^*\})}{Z^{\rm RMT}_{2N_f,\nu}(M_1, M_2)}, \qquad (4.45)
\end{aligned}
$$

where the last step relies on the observation that

$$
\begin{aligned}
&(\Lambda^2 - \Lambda_k^2)(\Lambda^{*2} - \Lambda_k^{*2}) \prod_{f=1}^{N_f} (m_f^2 - \Lambda_k^2)(\check{m}_f^2 - \Lambda_k^{*2}) \\
&= \Big\{ \prod_{f=1}^{N_f+1} (m_f^2 - \Lambda_k^2)(\check{m}_f^2 - \Lambda_k^{*2}) \Big\} \Big|_{m_{N_f+1}=\Lambda,\ \check{m}_{N_f+1}=\Lambda^*} \qquad (4.46)
\end{aligned}
$$

is nothing but the fermion determinant for $2N_f + 2$ flavors. This completes the second step.

• **Step 3**: The next task is to express the partition function as a determinant. This can be achieved with the method of Kieburg and Guhr [24]. Applying their theorem to the integral (4.39) we readily obtain

$$
Z^{\rm RMT}_{2N_f,\nu}(M_1, M_2) \sim \frac{\det\limits_{1 \le f, g \le N_f} \Big[ Z^{\rm RMT}_{2,\nu}(m_f, \check{m}_g) \Big]}{\Delta_{N_f}(m^2) \Delta_{N_f}(\check{m}^2)}. \qquad (4.47)
$$

The overall multiplicative constant that does not depend on the masses is dropped.

• **Step 4**: Now we substitute (4.47) into (4.42). With a bit of algebra, we get

$$
R_{2N_f,\nu}(\Lambda) \sim \frac{1}{|\Lambda|^{2\nu}} w_\nu(\Lambda) \frac{\det\limits_{1 \le f, g \le N_f+1} \Big[ Z^{\rm RMT}_{2,\nu}(m_f, \check{m}_g) \Big] \Big|_{m_{N_f+1}=\Lambda,\ \check{m}_{N_f+1}=\Lambda^*}}{\det\limits_{1 \le f, g \le N_f} \Big[ Z^{\rm RMT}_{2,\nu}(m_f, \check{m}_g) \Big]} \qquad (4.48)
$$

The final task is to take the $N \to \infty$ limit to derive the universal microscopic limit of the spectral density

$$\rho_{2N_f,\nu}(\Lambda) \equiv \lim_{N\to\infty} \frac{1}{2N} R_{2N_f,\nu}\left(\frac{\Lambda}{\sqrt{2N}}\right) \tag{4.49}$$

$$= \lim_{N\to\infty} \left\langle \sum_{i=1}^{N} \delta^2(\Lambda - \sqrt{2N}\Lambda_i) \right\rangle \tag{4.50}$$

with $\boxed{\tau_f \equiv \sqrt{2N} m_f}$ and $\boxed{\check{\tau}_f \equiv \sqrt{2N} \check{m}_f}$ kept finite.

Using (4.37), we finally arrive at the main formula

$$\boxed{\rho_{2N_f,\nu}(\Lambda) = \frac{2}{\pi}|\Lambda|^2 K_\nu(|\Lambda|^2) \frac{\left.\det\limits_{1\le f,g\le N_f+1}\Big[I_\nu(\tau_f\check{\tau}_g)\Big]\right|_{\tau_{N_f+1}=\Lambda,\ \check{\tau}_{N_f+1}=\Lambda^*}}{\det\limits_{1\le f,g\le N_f}\Big[I_\nu(\tau_f\check{\tau}_g)\Big]}} \tag{4.51}$$

which has not been known before. The method outlined above may be straightforwardly extended to the computations of higher-point correlation functions. The overall normalization constant was fixed to $2/\pi$ by matching with ([25], Eq. (3.45)) (quenched result), namely we impose the condition that $\rho_{2N_f,\nu}(\Lambda) \to \frac{1}{\pi}$ as $|\Lambda| \to \infty$.

It is important to note that, unlike two-color QCD, there are no exactly real or purely imaginary eigenvalues in this ChRMT.

## Figures

The simplest case is obviously the case $2N_f = 2$, with masses $(\tau, \check{\tau})$:

$$\rho_{2,\nu}(\Lambda) = \frac{2}{\pi}|\Lambda|^2 K_\nu(|\Lambda|^2) \left\{ I_\nu(|\Lambda|^2) - \frac{I_\nu(\tau\Lambda^*) I_\nu(\check{\tau}\Lambda)}{I_\nu(\tau\check{\tau})} \right\}. \tag{4.52}$$

In the chiral limit ($\tau, \check{\tau} \to 0$), $\rho_{2,\nu}(\Lambda)$ is rotationally symmetric around the origin. This is not true for two-color QCD ($\beta = 1$) and is a distinctive feature of this ensemble ($\beta = 2$).

In Fig. 4.1 the plot of $\rho_{2,\nu}(\Lambda)$ for equal masses is shown. (The spectrum in the region $\mathrm{Re}\Lambda < 0$ follows via chiral symmetry $\Lambda \leftrightarrow -\Lambda$.) The depletion of eigenvalues near $\Lambda = \tau$ and near the origin is clearly visible. The latter effect is pronounced for $\nu = 10$. Unlike two-color QCD, there is no repulsion of eigenvalues from the real and imaginary axis.

In Fig. 4.2 we show $\rho_{2,\nu}(\Lambda)$ for unequal masses. The spectral "density" is no longer positive definite and a circular domain of oscillation appears between $\tau$ and $\check{\tau}$. This behavior was also seen in the spectrum of two-color QCD, see Sect. 3.5.

## 4.3 QCD with Real Quarks at Large Density ($\beta = 4$)

### *4.3.1 QCD Side*

The theory we consider is a gauge theory with $N_f$ Dirac fermions in a *real* representation of the gauge group (for example, the adjoint representation of SU($N_c$) and the vector representation of SO($N_c$)).
This theory differs from two-color QCD ($\beta = 1$) in several respects:

(i) A color-singlet fermion may be formed from a quark and a gauge boson.
(ii) A color singlet diquark is symmetric under the exchange of two color indices, in contrast to two-color QCD where the indices are antisymmetric. This leads to a different pattern of global symmetry breaking.
(iii) The Dirac operator in this class of theories has a special anti-unitary symmetry

$$[\gamma_5 CK, \mathcal{D}(\mu)] = 0 \quad \text{with} \quad (\gamma_5 CK)^2 = -1, \tag{4.53}$$

where $K$ is the operator of complex conjugation. Consequently all nonzero eigenvalues of $\mathcal{D}(\mu)$ appear in quadruplets $(\lambda_n, -\lambda_n, \lambda_n^*, -\lambda_n^*)$. The eigenvector for $\lambda_n$ and that for $\pm\lambda_n^*$ are proven to be linearly independent [26]. This forbids the existence of exactly real or purely imaginary eigenvalues, in contrast to two-color QCD.
(iv) The quadruplet structure of the Dirac spectrum implies that the quark determinant (for nonzero eigenvalues) is strictly positive:

$$\det{}' \left(\mathcal{D}(\mu) + m\right) = \prod_n \left|\lambda_n^2 - m^2\right|^2 > 0\,. \tag{4.54}$$

Therefore this class of theories is free from the sign problem for any $N_f$ and for arbitrary masses. This makes contrast with two-color QCD where even $N_f$ and pairwise degenerate masses are necessary to have positive measure.

The most popular theory in this class seems to be the SU($N_c$) gauge theory with adjoint quarks. There is a vast literature on this subject and it is impossible to cite all articles; see [7, 22, 26–30] and references therein. Recently SO($N_c$) gauge theories at finite density has also attracted interests, in relation to the so-called orbifold equivalence at large $N_c$ [12, 31, 32].

### Ground State and Symmetry Breaking at $\mu \gg \Lambda$

Because a real representation is equivalent to its conjugate representation, a left-handed spinor and the complex conjugate of a right-handed spinor can be arranged into a single extended multiplet, as in two-color QCD (Sect. 2.2.1). Consequently the global symmetry of the Lagrangian at $\mu = 0$ in the chiral limit is enhanced to $\mathrm{U}(2N_f)$ for $N_f$ Dirac fermions [28]. Adding nonzero $\mu$ breaks this symmetry explicitly down to the conventional group

$$G = \mathrm{U}(N_f)_R \times \mathrm{U}(N_f)_L. \tag{4.55}$$

At sufficiently high density, the $\mathrm{U}(1)_A$ anomaly may be neglected. The quasi-free quarks will form Fermi surfaces and the attractive interaction in the color-singlet diquark channel drives the system toward the condensation of Cooper pairs according to the BCS mechanism:

$$\left\langle \psi_a^{fT} C\gamma_5 E^{ab} \psi_b^g \right\rangle \sim \frac{\mu^2 \Delta}{g} \delta^{fg}, \tag{4.56}$$

where $E$ is a symmetric matrix in color space that is specific to the given real representation of quarks. The color superconductivity does not take place; the system is a BCS-type superfluid in the confining phase. Since the diquark pairing takes place with the *same* flavor, this condensate will be sustained even for non-degenerate masses. Thus, the ground state is only invariant under the smaller group

$$H = \mathrm{SO}(N_f)_R \times \mathrm{SO}(N_f)_L. \tag{4.57}$$

The SSB $G \to H$ gives rise to $N_f^2 + N_f$ NG modes, which dominate the low-energy physics near the Fermi surface.

Since the measure is positive definite for any $\mu$ the method of QCD inequalities [15, 16] can be applied. One can show that the symmetry breaking pattern (4.56) is indeed consistent with the constraints from QCD inequalities.

### Chiral Lagrangian at High Density

The low-energy effective theory can be constructed by means of the spurion analysis. The coset space on which the NG modes live is

$$G/H \cong \big(\mathrm{U}(N_f)/\mathrm{SO}(N_f)\big)_1 \times \big(\mathrm{U}(N_f)/\mathrm{SO}(N_f)\big)_2. \tag{4.58}$$

The elements of $\big(\mathrm{U}(N_f)/\mathrm{SO}(N_f)\big)_i$ can be parametrized as $\Sigma_i = U_i U_i^T$ $(i = 1, 2)$ with $U_i \in \mathrm{U}(N_f)$ [27]. They are related to the quark fields as

$$\Sigma_1^{fg} \sim (\psi_R)_a^f C E^{ab} (\psi_R^T)_b^g, \quad \Sigma_2^{fg} \sim (\psi_L)_a^f C E^{ab} (\psi_L^T)_b^g. \tag{4.59}$$

Under a global unitary transformation

$$\psi_R \to g_R \psi_R \tag{4.60}$$
$$\psi_L \to g_L \psi_L \tag{4.61}$$

the coset elements transform as

$$\Sigma_1 \to g_R \Sigma_1 g_R^T, \quad \Sigma_2 \to g_L \Sigma_2 g_L^T \tag{4.62}$$

while the mass matrix must transform as

$$M \to g_L M g_R^\dagger \tag{4.63}$$

to keep the mass term in the Lagrangian $\bar{\psi}_L M \psi_R + \bar{\psi}_R M^\dagger \psi_L$ invariant.

Therefore the static part of the effective Lagrangian at $\mathcal{O}(M^2)$ is uniquely determined to be

$$\mathcal{L}_{\text{stat}} = -\Lambda^2 \text{tr}[M \Sigma_1 M^T \Sigma_2^\dagger] + \text{c.c.} \tag{4.64}$$

where $\Lambda$ is a low energy constant of mass dimension 1. The sign was chosen so that $\Sigma_1 = \Sigma_2 = \mathbf{1}$ is the ground state. There is no invariant term at $\mathcal{O}(M)$.

The constant term tr$[M^\dagger M]$ is neglected as it does not affect the dynamics of the NG modes (see also the Appendix A.3. The effective chemical potential of $\mathcal{O}(M^2)$ (the so-called Bedaque-Schäfer term in the CFL phase of three-color QCD [18]) is also dropped since it is suppressed by $1/\mu$ at large $\mu$.

### Matching with HDET

We now show that the low energy constant $\Lambda$ in $\mathcal{L}_{\text{stat}}$ is equal to $\Delta$ (the BCS gap) up to a numerical factor, through the matching with HDET [19, 20]. The procedure is just a rerun of arguments for $\beta = 1$ and 2. As noted in the Appendix A.2 the mass term induces the following four-fermion vertex in HDET [21]:

$$\begin{aligned}\mathcal{L}_{\text{mass}} = &\frac{g^2}{8\mu^4} (\psi_{i,L}^{\dagger\,a} C \psi_{j,L}^{*\,b})(\psi_{k,R}^{T\,c} C \psi_{l,R}^d) \\ &\times \left[(T^A)_{ac}(T^A)_{bd} M_{ik} M_{jl}\right] + (L \leftrightarrow R,\ M \leftrightarrow M^\dagger),\end{aligned} \tag{4.65}$$

where $i, j, k, l$ $(a, b, c, d)$ are flavor (color) indices. The color generators are in a *real* representation of the gauge group. Plugging (4.56) into this formula we find the shift of the ground state energy density $\delta\mathcal{E}$ due to the mass term to be

$$\delta\mathcal{E} \propto -\frac{g^2}{\mu^4}\left(\frac{\mu^2\Delta}{g}\right)^2 \mathrm{tr}[M^T M] + \text{c.c.} \tag{4.66}$$

$$= -\Delta^2 \mathrm{tr}[M^T M] + \text{c.c.} \tag{4.67}$$

Matching with (4.64) yields $\Lambda \sim \Delta$, as claimed.

### The $\varepsilon$-Regime

The $\varepsilon$-regime (or the microscopic domain) for this class of theories can be defined as in the cases of $\beta = 1$ and 2. The masses of the NG modes $m_{\mathrm{NG}}$ are of order $m_f \Delta/\mu$ (cf. Eq. (3.36)), whereas the mass of the lightest non-NG particle is $\sim \Delta$. If the system is put in a four-dimensional Euclidean box of size $L \times L \times L \times \beta \sim L^4$ with $\beta = 1/T$ satisfying

$$\frac{1}{\Delta} \ll L \ll \frac{1}{m_{\mathrm{NG}}}, \tag{4.68}$$

then the zero-momentum modes of the NG modes will dominate the partition function. In this domain, the kinetic term of the effective theory drops out and the finite-volume partition function is simply given by

$$Z_{N_f}^{\mathrm{QCD}}(M) = \int\limits_{\mathrm{U}(N_f)} [dU_1] \int\limits_{\mathrm{U}(N_f)} [dU_2] \exp\left(\# V_4 \Delta^2 \,\mathrm{Re}\,\mathrm{tr}\big[M U_1 U_1^T M^T (U_2 U_2^T)^\dagger\big]\right). \tag{4.69}$$

## 4.3.2 RMT Side

The ChRMT we consider is defined by

$$Z_{N_f,\bar{\nu}}^{\mathrm{RMT}}(M) = \int\limits_{\mathbb{H}^{N(N+\bar{\nu})}} [dV] \int\limits_{\mathbb{H}^{N(N+\bar{\nu})}} [dW] \prod_{f=1}^{N_f} \det\begin{pmatrix} m_f^* \mathbf{1}_N & V \\ W^\dagger & m_f \mathbf{1}_{N+\bar{\nu}} \end{pmatrix} \mathrm{e}^{-N\mathrm{tr}[VV^\dagger + WW^\dagger]} \tag{4.70}$$

where $V$ and $W$ are $N \times (N + \bar{\nu})$ matrices whose elements are real quaternions ($\beta = 4$). The integer $\bar{\nu}$ is arbitrary at this stage.

The determinant is invariant under $\mathrm{U}(N_f) \times \mathrm{U}(N_f)$. To check the symmetry breaking pattern of this model, we take the chiral limit and rewrite the fermion part as follows:

$$\prod_{f=1}^{N_f} \det \begin{pmatrix} \mathbf{0} & V \\ W^\dagger & \mathbf{0} \end{pmatrix} = \det{}^{N_f} V \cdot \det{}^{N_f} W \tag{4.71}$$

$$= \mathrm{Qdet}^{N_f} \begin{pmatrix} \mathbf{0} & V \\ V^\dagger & \mathbf{0} \end{pmatrix} \mathrm{Qdet}^{N_f} \begin{pmatrix} \mathbf{0} & W \\ W^\dagger & \mathbf{0} \end{pmatrix}, \tag{4.72}$$

where Qdet stands for the quaternion determinant [33]. The factor involving $V$ is exactly the fermion determinant in $\beta = 4$ ChRMT for QCD with $N_f$ Majorana fermions at zero density [34], and likewise for the second factor involving $W$. Therefore the pattern of SSB at large $N$ must be [34]

$$\mathrm{U}(N_f) \times \mathrm{U}(N_f) \to \mathrm{SO}(N_f) \times \mathrm{SO}(N_f)\,. \tag{4.73}$$

This coincides with the SSB $G \to H$ on the QCD side. This observation offers a quick check that this model should be the right ChRMT for dense QCD with real quarks.

In what follows, we use the basis $\boxed{\tau_\mu := (\mathbf{1}, -i\vec{\sigma})}$ for quaternions and view $V$ and $W$ as ordinary $2N \times 2(N+\bar{\nu})$ matrices with complex entries. Then the "det" and "tr" in (4.70) are to be interpreted as the determinant and the trace of such extended matrices.

### The $\sigma$-Model Representation: Matching with QCD

We shall derive the asymptotic form of $Z^{\mathrm{RMT}}$ in the large-$N$ limit. In the quaternion basis, the elements of $V$ and $W$ may be written as

$$V_{ij} = A^\mu_{ij}\tau_\mu \quad \text{and} \qquad W_{ij} = B^\mu_{ij}\tau_\mu\,, \tag{4.74}$$

where $A$ and $B$ are $N \times (N+\bar{\nu})$ *real* matrices. Then

$$Z^{\mathrm{RMT}}_{N_f,\bar{\nu}}(M) = \int \prod_{\mu,i,j} dA^\mu_{ij} dB^\mu_{ij} \times \prod_{f=1}^{N_f} \det \begin{pmatrix} m_f^* \mathbf{1}_{2N} & A^\mu\tau_\mu \\ (B^\nu\tau_\nu)^\dagger & m_f \mathbf{1}_{2(N+\bar{\nu})} \end{pmatrix} \mathrm{e}^{-2N\mathrm{tr}[A^\mu A^{\mu T} + B^\mu B^{\mu T}]} \tag{4.75}$$

$$= \int \prod_{\mu,i,j} dA^\mu_{ij} dB^\mu_{ij}\ \mathrm{e}^{-2N\mathrm{tr}[A^\mu A^{\mu T} + B^\mu B^{\mu T}]} \int d[\psi_R, \psi_L, \bar{\psi}_R, \bar{\psi}_L] \times \exp\left[ \begin{pmatrix} \bar{\psi}_R \\ \bar{\psi}_L \end{pmatrix} \begin{pmatrix} M^\dagger \otimes \mathbf{1}_{2N} & \mathbf{1}_{N_f} \otimes A^\mu\tau_\mu \\ \mathbf{1}_{N_f} \otimes (B^\nu\tau_\nu)^\dagger & M \otimes \mathbf{1}_{2(N+\bar{\nu})} \end{pmatrix} \begin{pmatrix} \psi_L \\ \psi_R \end{pmatrix} \right]. \tag{4.76}$$

Aside from the mass term, the exponent consists of two pieces. The part containing $A^\mu$ reads

$$
\begin{aligned}
& -2N(A^\mu_{ij})^2 + \bar{\psi}^f_{Ri} A^\mu_{ij} \tau_\mu \psi^f_{Rj} \\
&= -2N\left(A^\mu_{ij} - \frac{1}{4N}\bar{\psi}^f_{Ri}\tau_\mu\psi^f_{Rj}\right)^2 + \frac{1}{8N}\bar{\psi}^f_{Ri}\tau_\mu\psi^f_{Rj}\bar{\psi}^g_{Ri}\tau_\mu\psi^g_{Rj}.
\end{aligned}
\tag{4.77}
$$

Similarly the part containing $B^\nu$ reads

$$
\begin{aligned}
& -2N(B^\nu_{ij})^2 + \bar{\psi}^f_{Li} B^\nu_{ji} \tau^\dagger_\nu \psi^f_{Lj} \\
&= -2N\left(B^\nu_{ji} - \frac{1}{4N}\bar{\psi}^f_{Li}\tau^\dagger_\nu\psi^f_{Lj}\right)^2 + \frac{1}{8N}\bar{\psi}^f_{Li}\tau^\dagger_\nu\psi^f_{Lj}\bar{\psi}^g_{Li}\tau^\dagger_\nu\psi^g_{Lj}.
\end{aligned}
\tag{4.78}
$$

After integrating out $A^\mu_{ij}$ and $B^\mu_{ij}$, we find for the partition function

$$
\begin{aligned}
Z^{\rm RMT}_{N_f,\bar{\nu}}(M) = \int d[\psi_R, \psi_L, \bar{\psi}_R, \bar{\psi}_L] \exp\Big[&\bar{\psi}^f_{Ri} M^\dagger_{fg}\psi^g_{Li} + \bar{\psi}^f_{Li} M_{fg}\psi^g_{Ri} \\
&+ \frac{1}{8N}\bar{\psi}^f_{Ri}\tau_\mu\psi^f_{Rj}\bar{\psi}^g_{Ri}\tau_\mu\psi^g_{Rj} + \frac{1}{8N}\bar{\psi}^f_{Li}\tau^\dagger_\nu\psi^f_{Lj}\bar{\psi}^g_{Li}\tau^\dagger_\nu\psi^g_{Lj}\Big].
\end{aligned}
\tag{4.79}
$$

We denote the two components of fermion fields as

$$
\psi_{R/L} = \begin{pmatrix} \chi_{R/L} \\ \lambda_{R/L} \end{pmatrix}, \quad \bar{\psi}_{R/L} = \begin{pmatrix} \bar{\chi}_{R/L} \\ \bar{\lambda}_{R/L} \end{pmatrix}
\tag{4.80}
$$

and use the identity

$$
\sum_{\mu=0}^{3} (\tau_\mu)^{\alpha\beta}(\tau_\mu)^{\gamma\delta} = 2(\delta_{\alpha\beta}\delta_{\gamma\delta} - \delta_{\alpha\delta}\delta_{\beta\gamma}) = \sum_{\bar{\nu}=0}^{3} (\tau^\dagger_\nu)^{\alpha\beta}(\tau^\dagger_\nu)^{\gamma\delta}
\tag{4.81}
$$

to obtain

$$
\begin{aligned}
Z^{\rm RMT}_{N_f,\bar{\nu}}(M) = \int d[\lambda, \bar{\lambda}, \chi, \bar{\chi}] \exp\Big[&M_{fg}(\bar{\chi}^f_L\chi^g_R + \bar{\lambda}^f_L\lambda^g_R) + M^\dagger_{fg}(\bar{\chi}^f_R\chi^g_L + \bar{\lambda}^f_R\lambda^g_L) \\
&+ \frac{1}{4N}(\bar{\lambda}^f_R\bar{\chi}^g_R + \bar{\lambda}^g_R\bar{\chi}^f_R)(\chi^f_R\lambda^g_R + \chi^g_R\lambda^f_R) \\
&+ \frac{1}{4N}(\bar{\lambda}^f_L\bar{\chi}^g_L + \bar{\lambda}^g_L\bar{\chi}^f_L)(\chi^f_L\lambda^g_L + \chi^g_L\lambda^f_L)\Big].
\end{aligned}
\tag{4.82}
$$

The next task is to perform the Hubbard-Stratonovich transformation to bilinearize the four-fermion vertices. Introducing $N_f \times N_f$ complex *symmetric* matrices $\sigma_1$ and $\sigma_2$, we can rewrite the partition function as

$$
\begin{aligned}
Z^{\mathrm{RMT}}_{N_f,\bar{\nu}}(M) &= \int d[\sigma_1,\sigma_2]\ \mathrm{e}^{-N\mathrm{tr}[\sigma_1^\dagger\sigma_1+\sigma_2^\dagger\sigma_2]} \int d[\lambda,\bar{\lambda},\chi,\bar{\chi}] \\
&\quad \times \exp\Big[M_{fg}(\bar{\chi}_L^f\chi_R^g + \bar{\lambda}_L^f\lambda_R^g) + M^\dagger_{fg}(\bar{\chi}_R^f\chi_L^g + \bar{\lambda}_R^f\lambda_L^g) \\
&\quad +(\sigma_1)^{fg}\bar{\lambda}_R^f\bar{\chi}_R^g + (\sigma_1^\dagger)^{fg}\chi_R^f\lambda_R^g + (\sigma_2)^{fg}\bar{\lambda}_L^f\bar{\chi}_L^g + (\sigma_2^\dagger)^{fg}\chi_L^f\lambda_L^g\Big] \qquad (4.83) \\
&= \int d[\sigma_1,\sigma_2]\ \mathrm{e}^{-N\mathrm{tr}[\sigma_1^\dagger\sigma_1+\sigma_2^\dagger\sigma_2]} \int d[\lambda,\bar{\lambda},\chi,\bar{\chi}] \\
&\quad \times \exp\left[\frac{1}{2}\begin{pmatrix}\bar{\chi}_R\\ \bar{\lambda}_R\\ \chi_L\\ \lambda_L\\ \bar{\chi}_L\\ \bar{\lambda}_L\\ \chi_R\\ \lambda_R\end{pmatrix}^T
\begin{pmatrix}
\mathbf{0} & -\sigma_1 & M^\dagger & \mathbf{0} & & & & \\
\sigma_1 & \mathbf{0} & \mathbf{0} & M^\dagger & & & & \\
-M^* & \mathbf{0} & \mathbf{0} & \sigma_2^\dagger & & & & \\
\mathbf{0} & -M^* & -\sigma_2^\dagger & \mathbf{0} & & & & \\
 & & & & \mathbf{0} & -\sigma_2 & M & \mathbf{0} \\
 & & & & \sigma_2 & \mathbf{0} & \mathbf{0} & M \\
 & & & & -M^T & \mathbf{0} & \mathbf{0} & \sigma_1^\dagger \\
 & & & & \mathbf{0} & -M^T & -\sigma_1^\dagger & \mathbf{0}
\end{pmatrix}
\begin{pmatrix}\bar{\chi}_R\\ \bar{\lambda}_R\\ \chi_L\\ \lambda_L\\ \bar{\chi}_L\\ \bar{\lambda}_L\\ \chi_R\\ \lambda_R\end{pmatrix}\right]
\end{aligned}
\qquad (4.84)
$$

where the tensor products ($\otimes \mathbf{1}_N$ in the upper-left block and $\otimes \mathbf{1}_{N+\bar{\nu}}$ in the lower-right block) were omitted for brevity. The integration over the grassmann variables leaves us with pfaffians:

$$
\begin{aligned}
Z^{\mathrm{RMT}}_{N_f,\bar{\nu}}(M) &= \int d[\sigma_1,\sigma_2]\,\mathrm{e}^{-N\mathrm{tr}[\sigma_1^\dagger\sigma_1+\sigma_2^\dagger\sigma_2]}\,\mathrm{Pf}^N\begin{pmatrix}
\mathbf{0} & -\sigma_1 & M^\dagger & \mathbf{0} \\
\sigma_1 & \mathbf{0} & \mathbf{0} & M^\dagger \\
-M^* & \mathbf{0} & \mathbf{0} & \sigma_2^\dagger \\
\mathbf{0} & -M^* & -\sigma_2^\dagger & \mathbf{0}
\end{pmatrix} \\
&\quad \times \mathrm{Pf}^{N+\bar{\nu}}\begin{pmatrix}
\mathbf{0} & -\sigma_2 & M & \mathbf{0} \\
\sigma_2 & \mathbf{0} & \mathbf{0} & M \\
-M^T & \mathbf{0} & \mathbf{0} & \sigma_1^\dagger \\
\mathbf{0} & -M^T & -\sigma_1^\dagger & \mathbf{0}
\end{pmatrix}.
\end{aligned}
\qquad (4.85)
$$

To evaluate these pfaffians we use the identity

$$
\mathrm{Pf}\begin{pmatrix} A & B \\ -B^T & D\end{pmatrix} = \mathrm{Pf}\,D\cdot\mathrm{Pf}(A + BD^{-1}B^T) \qquad (4.86)
$$

with $A$ and $D$ antisymmetric and even-dimensional. This leads us to

$$
\begin{aligned}
Z^{\mathrm{RMT}}_{N_f,\bar{\nu}}(M) &= \int d[\sigma_1,\sigma_2]\ \mathrm{e}^{-N\mathrm{tr}[\sigma_1^\dagger\sigma_1+\sigma_2^\dagger\sigma_2]}(\det\sigma_2)^{N+\bar{\nu}}(\det\sigma_2^\dagger)^N \\
&\quad \times \det{}^{N+\bar{\nu}}\Big(\sigma_1^\dagger + M^T(\sigma_2)^{-1}M\Big)\det{}^N\Big(\sigma_1 + M^\dagger(\sigma_2^\dagger)^{-1}M^*\Big).
\end{aligned}
\qquad (4.87)
$$

All transformations so far are exact. Now let us take $N \to \infty$ keeping $\sqrt{N}\|M\| \sim \mathcal{O}(1)$. In this limit the integral may be evaluated with a saddle-point approximation. With a bit of algebra, we find that the saddle point manifold can be parametrized as

$$\sigma_1 = U_1 U_1^T \quad \text{and} \quad \sigma_2 = U_2 U_2^T \quad \text{with} \quad (U_1, U_2) \in \mathrm{U}(N_f) \times \mathrm{U}(N_f). \tag{4.88}$$

We then plug this into (4.87) and expand the determinant to the first nontrivial order, obtaining

$$\begin{aligned} Z_{N_f,\bar{\nu}}^{\mathrm{RMT}}(M) = &\int_{\mathrm{U}(N_f)} [dU_1] \int_{\mathrm{U}(N_f)} [dU_2] \det{}^{\bar{\nu}}(U_1^* U_1^\dagger U_2 U_2^T) \\ &\times \exp\left(2N \,\mathrm{Re}\,\mathrm{tr}[M U_1 U_1^T M^T U_2^* U_2^\dagger]\right). \end{aligned} \tag{4.89}$$

For $\bar{\nu} = 0$, this expression is exactly the same as the QCD partition function (4.69) in the $\varepsilon$-regime upon the identification (4.38). This ensures that all spectral sum rules are identical between QCD and ChRMT, which is a strong indication of the spectral universality: the spectral correlations of the Dirac operator in QCD may be computed exactly by means of ChRMT.

### Microscopic Spectral Density

The microscopic spectral correlation functions of this ChRMT have already been worked out in [35, 36] from a purely mathematical interest. The distinctive feature of the obtained spectral density, in comparison to $\beta = 1$ and 2, is that there is no oscillatory domain on the spectrum even for non-degenerate masses. This is natural in view of the absence of the sign problem in this model for arbitrary masses.

## References


1. G. Akemann, Matrix models and QCD with chemical potential. Int. J. Mod. Phys. **A22**, 1077–1122 (2007)
2. P. de Forcrand, Simulating QCD at finite density. PoS **LAT2009**, 010 (2009)
3. M.G. Alford, A. Kapustin, F. Wilczek, Imaginary chemical potential and finite fermion density on the lattice. Phys. Rev. **D59**, 054502 (1999)
4. P. de Forcrand, M.A. Stephanov, U. Wenger, On the phase diagram of QCD at finite isospin density. PoS **LAT2007**, 237 (2007)
5. D.T. Son, M.A. Stephanov, QCD at finite isospin density. Phys. Rev. Lett. **86**, 592–595 (2001)
6. D.T. Son, M.A. Stephanov, QCD at finite isospin density: from pion to quark antiquark condensation. Phys. Atom. Nucl. **64**, 834–842 (2001)
7. K. Splittorff, D.T. Son, M.A. Stephanov, QCD-like theories at finite baryon and isospin density. Phys. Rev. **D64**, 016003 (2001)
8. D. Toublan, J.B. Kogut, Isospin chemical potential and the QCD phase diagram at nonzero temperature and baryon chemical potential. Phys. Lett. **B564**, 212–216 (2003)
9. M.A. Metlitski, A.R. Zhitnitsky, Theta-dependence of QCD at finite isospin density. Phys. Lett. **B633**, 721–728 (2006)
10. J.M. Moller, On the phase diagram of QCD with small isospin chemical potential. Phys. Lett. **B683**, 235–239 (2010)

11. T. Schäfer, Patterns of symmetry breaking in QCD at high baryon density. Nucl. Phys. **B575**, 269–284 (2000)
12. M. Hanada, N. Yamamoto, Universality of phases in QCD and QCD-like theories. JHEP **02**, 138 (2012)
13. M.G. Alford, K. Rajagopal, F. Wilczek, Color-flavor locking and chiral symmetry breaking in high density QCD. Nucl. Phys. **B537**, 443–458 (1999)
14. R. Rapp, T. Schafer, E.V. Shuryak, M. Velkovsky, High-density QCD and instantons. Ann. Phys. **280**, 35–99 (2000)
15. D. Weingarten, Mass Inequalities for QCD. Phys. Rev. Lett. **51**, 1830 (1983)
16. E. Witten, Some inequalities among hadron masses. Phys. Rev. Lett. **51**, 2351 (1983)
17. R. Casalbuoni, G. Nardulli, Inhomogeneous superconductivity in condensed matter and QCD. Rev. Mod. Phys. **76**, 263–320 (2004)
18. P.F. Bedaque, T. Schafer, High density quark matter under stress. Nucl. Phys. **A697**, 802–822 (2002)
19. D.K. Hong, An effective field theory of QCD at high density. Phys. Lett. **B473**, 118–125 (2000)
20. D.K. Hong, Aspects of high density effective theory in QCD. Nucl. Phys. **B582**, 451–476 (2000)
21. T. Schafer, Mass terms in effective theories of high density quark matter. Phys. Rev. **D65**, 074006 (2002)
22. T. Kanazawa, T. Wettig, N. Yamamoto, Singular values of the dirac operator in dense QCD-like theories. JHEP **12**, 007 (2011)
23. J.C. Osborn, Universal results from an alternate random matrix model for QCD with a baryon chemical potential. Phys. Rev. Lett. **93**, 222001 (2004)
24. M. Kieburg, T. Guhr, Derivation of determinantal structures for random matrix ensembles in a new way. J. Phys. **A43**, 075201 (2010)
25. G. Akemann, M.J. Phillips, L. Shifrin, Gap Probabilities in non-hermitian random matrix theory. J. Math. Phys. **50**, 063504 (2009)
26. H. Leutwyler, A.V. Smilga, Spectrum of dirac operator and role of winding number in QCD. Phys. Rev. **D46**, 5607–5632 (1992)
27. A.V. Smilga, J.J.M. Verbaarschot, Spectral sum rules and finite volume partition function in gauge theories with real and pseudoreal fermions. Phys. Rev. **D51**, 829–837 (1995)
28. J.B. Kogut, M.A. Stephanov, D. Toublan, J.J.M. Verbaarschot, A. Zhitnitsky, QCD-like theories at finite baryon density. Nucl. Phys. **B582**, 477–513 (2000)
29. S. Hands et al., Numerical study of dense adjoint matter in two color QCD. Eur. Phys. J. **C17**, 285–302 (2000)
30. T. Zhang, T. Brauner, D.H. Rischke, QCD-like theories at nonzero temperature and density. JHEP **06**, 064 (2010)
31. A. Cherman, M. Hanada, D. Robles-Llana, Orbifold equivalence and the sign problem at finite baryon density. Phys. Rev. Lett. **106**, 091603 (2011)
32. A. Cherman, B.C. Tiburzi, Orbifold equivalence for finite density QCD and effective field theory. JHEP **06**, 034 (2011)
33. M.L. Mehta, *Random Matrices*, 3rd edn (Academic Press, Amsterdam, 2004)
34. A.M. Halasz, J.J.M. Verbaarschot, Effective lagrangians and chiral random matrix theory. Phys. Rev. **D52**, 2563–2573 (1995)
35. G. Akemann, The complex Laguerre symplectic ensemble of non-hermitian matrices. Nucl. Phys. **B730**, 253–299 (2005)
36. G. Akemann, F. Basile, Massive partition functions and complex eigenvalue correlations in matrix models with symplectic symmetry. Nucl. Phys. **B766**, 150–177 (2007)

# Chapter 5
# Conclusions

In this thesis, i extended the description of the Dirac spectrum in QCD vacuum in terms of the low energy effective theory of Nambu-Goldstone (NG) bosons and chiral random matrix theory (ChRMT), to the high density regime of a class of QCD-like theories, taking two-color QCD as a primary example. We discovered a novel correspondence between dense QCD and strongly non-Hermitian ChRMT, and thereby elucidated how the BCS pairing of quarks and the resulting symmetry breaking affect the distribution of small Dirac eigenvalues.

In Chap. 2 i have presented a quick introduction to the necessary background of this research, namely the phase structure of QCD at finite temperature density (Sect. 2.1), chiral perturbation theory (ChPT) in two-color QCD (Sect. 2.2) and ChRMT with chemical potential (Sect. 2.3), placing an emphasis on the universal correspondence between ChPT in the $\varepsilon$-regime and ChRMT in the large-$N$ limit of weak non-Hermiticity.

In Chap. 3 and Sect. 3.1 i discussed the $\varepsilon$-regime of the color-flavor-locked (CFL) phase, the ground state of QCD with three colors and three flavors at asymptotically high density. The comparison between an effective low-energy finite-volume partition function and its microscopic expression in terms of Dirac eigenvalues, enabled us to derive new Leutwyler-Smilga-type spectral sum rules of Dirac eigenvalues at high density. These sum rules indicate that the BCS gap $\Delta$ of quarks is a typical scale of statistical fluctuation of Dirac eigenvalues at high density, which makes an interesting contrast to the QCD vacuum where it is the chiral condensate $\langle \bar{q}q \rangle$ that sets the scale of Dirac eigenvalues. In Chap. 3 and Sect. 3.2 we moved on to two-color QCD and derived the effective theory of NG modes at high density, where the chiral symmetry is spontaneously broken by a gauge invariant diquark condensate $\langle qq \rangle$. We then defined "$\varepsilon$-regime" in two-color QCD at high density and derived nontrivial spectral sum rules for the Dirac eigenvalues. Our motivation to look into two-color QCD originates from the fact that the notorious sign problem of QCD at finite quark density, which makes conventional Monte Carlo techniques ineffective, can be avoided in two-color QCD if the number of flavors is even and the quark masses are pairwise degenerate. Our spectral sum rules therefore provide a new way

T. Kanazawa, *Dirac Spectra in Dense QCD*, Springer Theses,
DOI: 10.1007/978-4-431-54165-3_5, © Springer Japan 2013

to extract the BCS gap from the Dirac eigenvalue distribution measured in lattice simulations. We also defined the microscopic spectral density of Dirac eigenvalues in dense two-color QCD and argued, based on the obtained spectral sum rules, that it should be universal, i.e., solely determined by global symmetries while entirely insensitive to the ultraviolet dynamics of QCD.

To substantiate this claim, in Chap. 3 and Sect. 3.3 we introduced a strongly non-Hermitian ChRMT and showed that it reproduces all spectral sum rules of Dirac eigenvalues derived in Chap. 3 and Sect. 3.2. In Chap. 3 and Sect. 3.4 we carried out an exact computation of the microscopic spectral density of the proposed ChRMT for even number of flavors, **(I)** in the large-$N$ limit of weak non-Hermiticity, which is exactly mapped to two-color QCD at small chemical potential, as well as **(II)** in the large-$N$ limit of maximal non-Hermiticity, which is exactly mapped to the high-density BCS superfluid phase of two-color QCD. We found that a strong oscillation emerges in the spectral density in both limits when the sign problem is harsh. A similar phenomenon has also been observed in *three-color* QCD at *small* chemical potential [1, 2].

In Chap. 3 and Sect. 3.5 we evaluated the severity of the sign problem of dense two-color QCD using ChRMT in two ways. First, we approximated the exact formula of the microscopic spectral density of complex eigenvalues for $N_f = 2$ at maximal non-Hermiticity by leading exponential terms and showed that there are indeed two regions on the spectrum separated by a circle, outside which the density is flat and smooth while outside which the density oscillates with a small wavelength and a large amplitude. Moreover we argued that the envelope of the oscillatory domain can be interpreted as a line of first order phase transition in the theory *with two additional flavors*. Secondly, we computed the *average sign* of the fermion determinant exactly, as a function of $\mu$ in the limit of weak non-Hermiticity, and as a function of $|m_1 - m_2|$ in the limit of maximal non-Hermiticity, with $m_{1,2}$ the quark masses. In the former case, we found that the sign problem for $N_f = 1$ gets harsh when $\mu \gtrsim m_\pi/2$. In the latter case, we found that the sign problem for $N_f = 2$ worsens when $|m_1 - m_2| \gtrsim 4.5/\sqrt{V_4\Delta^2}$. In both cases we observed that the sign problem is milder in the sector of larger topological charge, as independently found in [3] at small chemical potential in three-color QCD. The novelty of our analysis is that we are the first to study the sign problem *in the BCS regime of dense QCD*.

Finally, we showed in Chap. 4 that our analysis on two-color QCD can be generalized to three classes of dense QCD-like theories: QCD with *pseudoreal* quarks at high baryon density ($\beta = 1$), QCD with complex quarks at high *isospin* density ($\beta = 2$), and QCD with *real* quarks at high baryon density ($\beta = 4$), where the words (pseudoreal/complex/real) refer to the property of the color representations of quarks, whereas $\beta$ stands for the usual Dyson index. Thus the "three-fold way" of ChRMT discovered by Verbaarschot in QCD vacuum [4] has now been extended to a drastically different regime, i.e., the weak-coupling BCS regime of high density QCD.

As is the case with every scientific research, our work poses various future challenges. One may ask **(I):** "If we have two distinct low-energy effective theories for two-color QCD at low density and at high density, how can they be interpolated at moderate density? Should we expect a phase transition in between?" And **(II):** "In two-color QCD we can, in principle, add two kinds of masses: the ordinary Dirac mass and the *Majorana mass* (i.e., the diquark source). What can we learn by adding the latter?" These questions were investigated in [5] and were answered completely. Regarding **(I)**, we found that there are three effective theories for two-color QCD at nonzero density. Each theory has its own domain of validity, and the overlap of domains of validity naturally accounts for a crossover from one theory to another as we increase the density. Thus a phase transition is not mandatory. As for **(II)**, the Majorana masses were found to be a probe of the *singular value spectrum* of the Dirac operator. While the eigenvalues are related to the BCS gap $\Delta$, the *singular values* of the Dirac operator are related to the diquark condensate $\langle qq \rangle$. The reader can find 'further details in [5].

The extension of our theoretical framework to the realistic case of three-color QCD at high baryon density is an important unsolved problem. One of the hallmarks of this theory is that the SU(3) gauge symmetry is completely broken via the Anderson-Higgs mechanism owing to the condensation of color-antitriplet diquarks. There are several difficulties to be solved in the application of RMT to this phase; see the discussion at the end of Chap. 3 Sect. 3.1. We defer this issue to future work. A direct check of our analytical results via first-principle lattice QCD simulations is another important direction to be pursued in the future.

## References


1. J.C. Osborn, K. Splittorff, J.J.M. Verbaarschot, Chiral symmetry breaking and the Dirac spectrum at nonzero chemical potential. Phys. Rev. Lett. **94**, 202001 (2005)
2. G. Akemann, J.C. Osborn, K. Splittorff, J.J.M. Verbaarschot, Unquenched QCD Dirac operator spectra at nonzero baryon chemical potential. Nucl. Phys. **B712**, 287–324 (2005)
3. J.C.R. Bloch, T. Wettig, Random matrix analysis of the QCD sign problem for general topology. JHEP **03**, 100 (2009)
4. J.J.M. Verbaarschot, The Spectrum of the QCD Dirac operator and chiral random matrix theory: the threefold way. Phys. Rev. Lett. **72**, 2531–2533 (1994)
5. T. Kanazawa, T. Wettig, N. Yamamoto, Singular values of the Dirac operator in dense QCD-like theories. JHEP **12**, 007 (2011)

# Appendix A

## A.1 Convention

First, the Pauli matrices are defined as

$$\sigma_1 = \begin{pmatrix} 0 & 1 \\ 1 & 0 \end{pmatrix}, \sigma_2 = \begin{pmatrix} 0 & -i \\ i & 0 \end{pmatrix}, \sigma_3 = \begin{pmatrix} 1 & 0 \\ 0 & -1 \end{pmatrix}. \tag{A.1}$$

In the four-dimensional *Minkowski* spacetime, we define

$$g_{\mu\nu} = \mathrm{diag}(1, -1, -1, -1), \qquad \mu, \nu = 0, 1, 2, 3 \tag{A.2}$$

$$\{\gamma^\mu, \gamma^\nu\} = 2g^{\mu\nu}, (\gamma^0)^\dagger = \gamma^0, (\gamma^i)^\dagger = -\gamma^i \tag{A.3}$$

$$\gamma_5 = i\gamma^0\gamma^1\gamma^2\gamma^3 \tag{A.4}$$

$$\gamma^0 = \begin{pmatrix} \mathbf{0} & \mathbf{1} \\ \mathbf{1} & \mathbf{0} \end{pmatrix}, \gamma^i = \begin{pmatrix} \mathbf{0} & -\sigma_i \\ \sigma_i & \mathbf{0} \end{pmatrix}, \gamma_5 = \begin{pmatrix} \mathbf{1} & \mathbf{0} \\ \mathbf{0} & -\mathbf{1} \end{pmatrix} \tag{A.5}$$

$$C = i\gamma^2\gamma^0 : \text{ The charge conjugation matrix} \tag{A.6}$$

$$C = C^* = -C^T = -C^{-1}, \qquad C^\dagger C = \mathbf{1} \tag{A.7}$$

$$C\gamma^\mu C = (\gamma^\mu)^T \tag{A.8}$$

$$\mathcal{L}_{\mathrm{QCD}} = \sum_{f=1}^{N_f} \overline{\psi}_f(i\gamma^\mu D_\mu - m_f + \mu\gamma^0)\psi_f - \frac{1}{4}F^a_{\mu\nu}F^{a\mu\nu} \tag{A.9}$$

$$\overline{\psi} = \psi^\dagger\gamma^0, \quad D_\mu = \partial_\mu + igA^a_\mu T^a, a = 1, 2, \ldots, N_c^2 - 1 \tag{A.10}$$

In the four-dimensional *Euclidean* spacetime, we define

$$x_0 = -ix_0^{\mathrm{E}}, \quad A_0^a = i(A^{\mathrm{E}})^a_0, \quad \gamma^i = -\gamma_i = i\gamma_i^{\mathrm{E}}, \tag{A.11}$$

T. Kanazawa, *Dirac Spectra in Dense QCD*, Springer Theses,
DOI: 10.1007/978-4-431-54165-3, © Springer Japan 2013

where the subscript 'E' for Euclidean quantities will be omitted below for simplicity. Then

$$\{\gamma_\mu, \gamma_\nu\} = 2\delta_{\mu\nu}, (\gamma_\mu)^\dagger = \gamma_\mu \tag{A.12}$$

$$\gamma_0 = \begin{pmatrix} \mathbf{0} & \mathbf{1} \\ \mathbf{1} & \mathbf{0} \end{pmatrix}, \gamma_i = \begin{pmatrix} \mathbf{0} & -i\sigma_i \\ i\sigma_i & \mathbf{0} \end{pmatrix}, \gamma_5 = \begin{pmatrix} \mathbf{1} & \mathbf{0} \\ \mathbf{0} & -\mathbf{1} \end{pmatrix} \tag{A.13}$$

$$C = i\gamma_0\gamma_2 = \begin{pmatrix} -\sigma_2 & \mathbf{0} \\ \mathbf{0} & \sigma_2 \end{pmatrix}, \quad C = -C^* = -C^T = C^{-1}, \qquad C^\dagger C = \mathbf{1} \tag{A.14}$$

$$C\gamma_\mu C = -(\gamma_\mu)^T = -\gamma_\mu^* \tag{A.15}$$

$$\mathcal{L}_{\mathrm{QCD}} = \sum_{f=1}^{N_f} \overline{\psi}_f(\gamma_\mu D_\mu + m_f - \mu\gamma_0)\psi_f + \frac{1}{4}F^a_{\mu\nu}F^a_{\mu\nu} \tag{A.16}$$

$$\overline{\psi} = \psi^\dagger\gamma_0, \quad D_\mu = \partial_\mu + igA^a_\mu T^a, a = 1, 2, \ldots, N_c^2 - 1 \tag{A.17}$$

The expressions for $\gamma_0$ and $\gamma_i$ can be unified to

$$\gamma_\mu = \begin{pmatrix} \mathbf{0} & -i\sigma_\mu \\ i\sigma_\mu^\dagger & \mathbf{0} \end{pmatrix} \qquad \text{for } \sigma_\mu = (i\mathbf{1}, \sigma_1, \sigma_2, \sigma_3)\,. \tag{A.18}$$

Note that the Dirac operator $\gamma_\mu D_\mu$ is *anti-Hermitian* in our convention. With a nonzero chemical potential, the operator $\mathcal{D}(\mu) \equiv \gamma_\mu D_\mu - \mu\gamma_0$ is no longer anti-Hermitian. Instead it satisfies

$$\mathcal{D}(\mu)^\dagger = -\mathcal{D}(-\mu)\,. \tag{A.19}$$

## A.2 Microscopic Derivation of Mass Terms in the Effective Theory

In this appendix we give the detailed derivation of the coefficient of the mass term in the effective Lagrangian at high density, (3.4) and (3.29). For this purpose, we calculate the shift of the vacuum energy due to the quark mass from the microscopic theory (QCD), which should be matched against the result obtained from the low-energy effective Lagrangian. The starting point is the mass term of the microscopic theory given in [1],

$$\mathcal{L}_{\mathrm{mass}} = \frac{g^2}{8\mu^4}(\psi^{\dagger\,a}_{i,L} C\psi^{*\,b}_{j,L})(\psi^{T\,c}_{k,R} C\psi^d_{l,R})\left[(T^A)^{ac}(T^A)^{bd} M_{ik} M_{jl}\right] + (L \leftrightarrow R,\, M \leftrightarrow M^\dagger)\,, \tag{A.20}$$

where $i,j,k,l(a,b,c,d)$ are flavor (color) indices and $T^A$ ($A = 1,\ldots,N_c^2 - 1$) are $SU(N_c)$ color generators with normalization $\mathrm{tr}(T^A T^B) = \delta^{AB}/2$ . Let us focus on the case $N_c = 2$ for concreteness. The diquark condensates $d_L$ and $d_R$ that are antisymmetric in flavor and color can be written as

$$\langle \psi_{i,L}^{T\,a} C \psi_{j,L}^b \rangle = \varepsilon^{ab} I_{ij} d_L \,, \qquad \langle \psi_{i,R}^{T\,a} C \psi_{j,R}^b \rangle = \varepsilon^{ab} I_{ij} d_R \,, \tag{A.21}$$

where $I$ is defined in (2.17). The expectation values of $d_L$ and $d_R$ in the ground state can be evaluated in weak coupling. Generalizing the result of [1] to arbitrary $N_c$, we obtain

$$|d_L| = |d_R| = \sqrt{\frac{6N_c}{N_c+1} \frac{\mu^2 \Delta}{\pi g}} \,. \tag{A.22}$$

Using (A.21), (A.22), and the Fierz identity

$$(T^A)^{ac}(T^A)^{bd} = \frac{1}{2}\delta^{ad}\delta^{bc} - \frac{1}{2N_c}\delta^{ac}\delta^{bd} \,, \tag{A.23}$$

the shift of the vacuum energy density due to the quark mass for $N_c = 2$ is found to be

$$\delta\mathcal{E} = \frac{3\Delta^2}{4\pi^2} \mathrm{tr}(MIM^T I) + (M \leftrightarrow M^\dagger) \,. \tag{A.24}$$

For $N_f = 2$ we have $I_{ij} = \varepsilon_{ij}$, and (A.24) reduces to

$$\delta\mathcal{E}_{N_f=2} = -\frac{3\Delta^2}{2\pi^2} \det M + (M \leftrightarrow M^\dagger) \,. \tag{A.25}$$

By comparing (A.24) or (A.25) with the vacuum energy obtained from the low-energy effective Lagrangian (with $\Sigma_L = \Sigma_R = I$) or (3.37), the values of the coefficients are found to be

$$c = \frac{3}{4\pi^2} \,, \qquad c' = \frac{3}{2\pi^2} \,, \tag{A.26}$$

respectively, where $c'$ corrects the value of $4/3\pi^2$ given in [2].

In the CFL phase for $N_f = N_c = 3$, (A.21) must be replaced by

$$\langle \psi_{i,H}^{T\,a} C \psi_{j,H}^b \rangle = (\delta_{ai}\delta_{bj} - \delta_{aj}\delta_{bi}) d_H \,, \qquad (H = L,\, R) \tag{A.27}$$

and the ensuing analysis goes along the same lines as for $N_c = 2$.

## A.3 tr$[MM^\dagger]$ in the ε-regime

As noted on Chap. 3 Sect. 3.1, the effective action of the CFL phase contains a term tr$[MM^\dagger]$, which was overlooked in [3].[1] This term is usually neglected in the literature as it is independent of the Nambu-Goldstone fields and irrelevant to dynamical properties of dense matter such as transport coefficients. However it matters when we are interested in the quark mass dependence of the partition function itself. The origin of this term is quite easy to understand: for a free quark, the shift of the energy due to a current mass is

$$\sqrt{\boldsymbol{p}^2 + m^2} - |\boldsymbol{p}| \approx \frac{m^2}{2|\boldsymbol{p}|}\,. \tag{A.28}$$

This contribution comes from all quarks inside the Fermi sphere of volume $\propto p_F^3 \sim \mu^3$. Therefore the total contribution would be of order

$$\mu^3 \times \frac{m^2}{2\mu} \sim \mu^2 m^2\,. \tag{A.29}$$

We can thus expect that there will be a term $C\mu^2 \mathrm{tr}[MM^\dagger]$ in the effective action at high density, with $C \sim \mathcal{O}(1)$ a numerical constant. Then (3.9) must be modified as

$$Z_{\mathrm{EFT}}(M) = \int_{\mathrm{U}(3)} [d\Sigma] \exp\left(\frac{3}{4\pi^2} V_4 \Delta^2 \left\{ (\mathrm{tr} M\Sigma^\dagger)^2 - \mathrm{tr}(M\Sigma^\dagger)^2 \right\} \det\Sigma + \mathrm{h.c.} + CV_4\mu^2 \mathrm{tr} MM^\dagger \right). \tag{A.30}$$

The expansion (3.11) must also be altered to

$$\begin{aligned} Z_{\mathrm{EFT}}(M) = 1 + CV_4\mu^2 \mathrm{tr}MM^\dagger + \frac{3}{8\pi^4}(V_4\Delta^2)^2 \left\{ (\mathrm{tr}MM^\dagger)^2 - \mathrm{tr}(MM^\dagger)^2 \right\} \\ + \frac{1}{2}(CV_4\mu^2)^2 (\mathrm{tr}MM^\dagger)^2 + \mathcal{O}(M^6)\,. \end{aligned} \tag{A.31}$$

Matching this expression with

$$\begin{aligned} Z_{\mathrm{QCD}}(M) = & 1 + \mathrm{tr}M^\dagger M \left\langle\!\!\left\langle {\sum_n}' \frac{1}{\lambda_n^2} \right\rangle\!\!\right\rangle + \frac{(\mathrm{tr}M^\dagger M)^2 - \mathrm{tr}(M^\dagger M)^2}{2} \left\langle\!\!\left\langle {\sum_n}' \frac{1}{\lambda_n^4} \right\rangle\!\!\right\rangle \\ & + (\mathrm{tr}M^\dagger M)^2 \left\langle\!\!\left\langle {\sum_{m\neq n}}' \frac{1}{\lambda_m^2 \lambda_n^2} \right\rangle\!\!\right\rangle + \mathcal{O}(M^6)\,, \end{aligned} \tag{A.32}$$

---

[1] Subleading terms like tr$[(M^\dagger M)^2]$ are negligibly small compared to the leading one.

yields the modified spectral sum rules (cf. (3.15) and (3.16)):

$$\left\langle\!\left\langle \sum_n{}' \frac{1}{\lambda_n^2} \right\rangle\!\right\rangle = CV_4\mu^2 \,, \tag{A.33}$$

$$\left\langle\!\left\langle \sum_n{}' \frac{1}{\lambda_n^4} \right\rangle\!\right\rangle = \frac{3}{4\pi^4}(V_4\Delta^2)^2 \,, \tag{A.34}$$

$$\left\langle\!\left\langle \sum_{m\neq n}{}' \frac{1}{\lambda_m^2\lambda_n^2} \right\rangle\!\right\rangle = \frac{1}{2}(CV_4\mu^2)^2 \,, \qquad \text{etc.} \tag{A.35}$$

The inclusion of $\mathrm{tr}[M^\dagger M]$ turned vanishing spectral sums into nonzero large values ($\leftarrow$ (A.33) and (A.35)). It is interesting, however, that (A.34) receives no correction from the bulk term $\mathrm{tr}[M^\dagger M]$, which guarantees that (3.17) is correct as it stands.

These modified sum rules involve two scales: $\mu$ and $\Delta$ with $\Delta \sim \mu e^{-1/g} \ll \mu$. While (A.33) suggests that the eigenvalues are $\mathcal{O}(1/\sqrt{V_4\mu^2})$, (A.34) implies that they are $\mathcal{O}(1/\sqrt{V_4\Delta^2})$. Which is the right scale?

This paradox can be resolved as follows. Let us start with the free Dirac operator $p - \mu\gamma^0$. (The omission of the gauge interaction makes sense as a rough approximation since the coupling constant is small at asymptotically high density owing to asymptotic freedom.) Since the momentum of those quarks that participate in the BCS pairing is typically $\mathcal{O}(\mu)$, it is in principle possible that the combination $\not{p} - \mu\gamma^0$ becomes much smaller than $\mu$ owing to a cancellation between the two terms. From the consistency with (A.34), the cancellation is expected to result in small Dirac eigenvalues of order $\mathcal{O}(1/\sqrt{V_4\Delta^2})$. It is this scale that arises from the BCS pairing at the Fermi surface.

On the other hand, quarks deep inside the Fermi sea do not participate in the collective motion of Cooper pairs. They carry arbitrary momentum from 0 to $\mu$, so that $\not{p} - \mu\gamma^0$ can take all values up to $\mathcal{O}(\mu)$. Therefore, schematically, we will get (in Euclidean spacetime)

$$\left\langle\!\left\langle \sum_n{}' \frac{1}{\lambda_n^2} \right\rangle\!\right\rangle \sim V_4 \int dp_0 \int_{|\boldsymbol{p}|<\mu} d^3\boldsymbol{p} \frac{1}{(p_0+i\mu)^2+\boldsymbol{p}^2} \sim V_4\mu^2 \,. \tag{A.36}$$

In this way we recover (A.33). We see that *it is the 4-dimensional integral* $\int d^4p$ *that transmutes small* $1/\lambda_n^2 \sim 1/\mu^2$ *into large* $V_4\mu^2$. Therefore it is erroneous to conclude from (A.33) that the scale of eigenvalues is $\mathcal{O}(1/\sqrt{V_4\mu^2})$; rather, the typical scale of eigenvalues associated with (A.33) is $\mathcal{O}(\mu)$.

It remains to explain why there is no contribution from 'hard' eigenvalues $\sim \mathcal{O}(\mu)$ to (A.34). Based on the same logic as above, we can estimate their contribution to be $\sim V_4\mu^4 \times (1/\mu)^4 \sim V_4$. This is negligibly small compared to the RHS of (A.34) $\propto V_4^2$ and may be safely dropped. (I conjecture that the term $\mathrm{tr}[MM^\dagger]$ will not modify any of the nonzero spectral sum rules derived in the $\varepsilon$-regime. A proof remains to be given.) In this way the initial contradiction gets completely resolved. The analysis pertains to dense two-color QCD as well.

The bottom line of this appendix is that, as long as the Dirac eigenvalues of magnitude $\mathcal{O}(1/\sqrt{V_4\Delta^2})$ are concerned, the approach based on the effective theory without the constant term $\mathrm{tr}[MM^\dagger]$ can be justified. Equations (A.33) and (A.35) never signal the inconsistency of our approach.

## A.4 Sign-Quenched Partition Functions

In two-color QCD the fermion determinant has no complex phase but only a sign, so we shall introduce a "sign-quenched" partition function, in which $\det[\mathcal{D}(\mu)+m]$ is replaced by $|\det[\mathcal{D}(\mu)+m]|$, as a direct counterpart of the phase-quenched partition function in three-color QCD. The purpose of this appendix is to derive analytical expressions for such sign-quenched partition functions, first for finite $N$, and later in the limits of both weak and strong non-Hermiticity. The results in this appendix will be used in the main text for the analysis of the sign problem.

### *A.4.1 Main Result at Finite N*

Let us begin with a formula for the probability measure, with $z_j \equiv x_j + iy_j$ ,

$$Z^{\mathrm{RMT}(N)}_{N_f,\nu}(\hat{\mu},\{\tau_f\}) = c_N \prod_{f=1}^{N_f} \tau_f^\nu \prod_{k=1}^{N} \int_{\mathbb{C}} d^2 z_k w_\nu(\hat{\mu},z_k) |\Delta_N(\{z\})| \prod_{j=1}^{N}\prod_{f=1}^{N_f} (\tau_f^2 + z_j)$$
$$\times \sum_{n=0}^{[N/2]} \frac{1}{n!(N-2n)!} \Big( \prod_{i=1}^{n} \delta^2(z_{2i-1} - z_{2i}^*) \Big) \delta(y_{2n+1}) \cdots \delta(y_N) \,, \tag{A.37}$$

which is equivalent to (3.93) and (3.94) but is more convenient for our current purpose. Also we made the $N$-dependence of $Z^{\mathrm{RMT}}_{N_f,\nu}$ explicit on the LHS. The weight function $w_\nu(\hat{\mu},z)$ is defined in (3.95). The prefactor $c_N$ is given in (3.98). We also define a "partially sign-quenched" partition function, which we denote by

$$Z^{\mathrm{RMT}(N)}_{\|N'_f\|+N_f,\nu}\left(\hat{\mu},\{\tilde{\tau}_k\}_{k=1}^{N'_f},\{\tau_\ell\}_{\ell=1}^{N_f}\right), \tag{A.38}$$

as being identical to $Z^{\mathrm{RMT}(N)}_{N'_f+N_f,\nu}$ in (A.37) except that the fermion determinant is replaced by

$$\prod_{j=1}^{N}\left\{\prod_{h=1}^{N'_f}|\tilde{\tau}_h|^\nu|\tilde{\tau}_h^2+z_j|\prod_{f=1}^{N_f}\tau_f^\nu(\tau_f^2+z_j)\right\}. \tag{A.39}$$

After these prerequisites, we can now state our main result at finite $N$:

$$R^{\mathbb{R}}_{N_f,\nu}(\hat{\mu};x_1,\ldots,x_k)$$
$$=\frac{c_N}{c_{N-k}}|\Delta_k(\{x\})|\prod_{i=1}^{k}\frac{w_\nu(\hat{\mu},x_i)}{|x_i|^{\nu/2}}\prod_{f=1}^{N_f}(\tau_f^2+x_i)\frac{Z^{\mathrm{RMT}(N-k)}_{\|k\|+N_f,\nu}(\hat{\mu},\{\sqrt{-x}\},\{\tau\})}{Z^{\mathrm{RMT}(N)}_{N_f,\nu}(\hat{\mu},\{\tau\})}. \tag{A.40}$$

That is, one can calculate a partially sign-quenched partition function from the $k$-point correlation function of *real* squared Dirac eigenvalues (and vice versa). Below we give a proof of this formula for $k=1$. The generalization to $k>1$ is then straightforward.

### A.4.2 Proof: The k = 1 Case

By definition, we have

$$R_{N_f,\nu}(\hat{\mu};z)=\frac{c_N}{Z^{\mathrm{RMT}(N)}_{N_f,\nu}(\hat{\mu},\{\tau\})}\prod_{f=1}^{N_f}\tau_f^\nu\prod_{k=1}^{N}\int_{\mathbb{C}}d^2z_k w_\nu(\hat{\mu},z_k)|\Delta_N(\{z\})|\prod_{j=1}^{N}\prod_{f=1}^{N_f}(\tau_f^2+z_j)\times$$
$$\times\sum_{n=0}^{[N/2]}\frac{1}{n!(N-2n)!}\prod_{i=1}^{n}\delta^2(z_{2i-1}-z_{2i}^*)\,\delta(y_{2n+1})\cdots\delta(y_N)\sum_{\ell=1}^{N}\delta^2(z-z_\ell). \tag{A.41}$$

For $2n+1\le\ell\le N, \delta^2(z-z_\ell)$ in the last sum will yield $\delta(y)$ (with $z\equiv x+iy$), hence

$$\begin{aligned}
\delta(y)R^{\mathbb{R}}_{N_f,\nu}(\hat{\mu};x) &= \frac{c_N}{Z^{\mathrm{RMT}(N)}_{N_f,\nu}(\hat{\mu},\{\tau\})}\prod_{f=1}^{N_f}\tau_f^{\nu}\prod_{k=1}^{N}\int_{\mathbb{C}} d^2z_k w_\nu(\hat{\mu},z_k)|\Delta_N(\{z\})|\prod_{j=1}^{N}\prod_{f=1}^{N_f}(\tau_f^2+z_j)\times\\
&\times\sum_{n=0}^{[N/2]}\frac{1}{n!(N-2n)!}\prod_{i=1}^{n}\delta^2(z_{2i-1}-z_{2i}^*)\,\delta(y_{2n+1})\cdots\delta(y_N)\sum_{\ell=2n+1}^{N}\delta^2(z-z_\ell)\\
&= \frac{c_N}{Z^{\mathrm{RMT}(N)}_{N_f,\nu}(\hat{\mu},\{\tau\})}\prod_{f=1}^{N_f}\tau_f^{\nu}\prod_{k=1}^{N}\int_{\mathbb{C}} d^2z_k w_\nu(\hat{\mu},z_k)|\Delta_N(\{z\})|\prod_{j=1}^{N}\prod_{f=1}^{N_f}(\tau_f^2+z_j)\times\\
&\times\sum_{n=0}^{[N/2]}\frac{N-2n}{n!(N-2n)!}\prod_{i=1}^{n}\delta^2(z_{2i-1}-z_{2i}^*)\,\delta(y_{2n+1})\cdots\delta(y_N)\,\delta^2(z-z_N)\\
&= \frac{c_N}{Z^{\mathrm{RMT}(N)}_{N_f,\nu}(\hat{\mu},\{\tau\})}w_\nu(\hat{\mu},x)\prod_{f=1}^{N_f}(\tau_f^2+x)\prod_{f=1}^{N_f}\tau_f^{\nu}\times\\
&\times\prod_{k=1}^{N-1}\int_{\mathbb{C}} d^2z_k w_\nu(\hat{\mu},z_k)\left(|\Delta_{N-1}(\{z\})|\prod_{\ell=1}^{N-1}|x-z_\ell|\right)\prod_{j=1}^{N-1}\prod_{f=1}^{N_f}(\tau_f^2+z_j)\times\\
&\times\sum_{n=0}^{[N/2]}\frac{1}{n!(N-1-2n)!}\prod_{i=1}^{n}\delta^2(z_{2i-1}-z_{2i}^*)\delta(y_{2n+1})\cdots\delta(y_{N-1})\delta(y)\,.
\end{aligned}\tag{A.42}$$

Therefore

$$R^{\mathbb{R}}_{N_f,\nu}(\hat{\mu};x) = \frac{c_N}{c_{N-1}}\frac{w_\nu(\hat{\mu},x)}{|x|^{\nu/2}}\prod_{f=1}^{N_f}(\tau_f^2+x)\frac{Z^{\mathrm{RMT}(N-1)}_{\|1\|+N_f,\nu}(\hat{\mu},\sqrt{-x},\{\tau\})}{Z^{\mathrm{RMT}(N)}_{N_f,\nu}(\hat{\mu},\{\tau\})}\,. \tag{A.43}$$

### *A.4.3 The Large-*N *Limit of Weak Non-Hermiticity*

Here we shall compute

$$\mathcal{Z}_w^{(N_f=\|1\|,\nu)}(\gamma,\kappa) \equiv \lim_{N\to\infty,\,\mathrm{weak}}\left\langle|\det[\mathcal{D}(\hat{\mu})+\tau]|\right\rangle_{N_f=0}, \tag{A.44}$$

where the limit stands for the large-$N$ limit of weak non-Hermiticity (cf. Chap. 3 Sect. 3.4.4), with

$$\begin{aligned}
\gamma &\equiv \sqrt{2N}\hat{\mu}\,,\\
\kappa_f &\equiv 2\sqrt{N}\tau_f\,,\\
\xi_k &\equiv 2\sqrt{N}\Lambda_k\,.
\end{aligned}$$

Putting $N_f = 0$ and $x = -\tau^2 \in \mathbb{R}$ in (A.43), we find

$$Z_{N_f=\|1\|,\nu}^{\mathrm{RMT}(N-1)}(\hat{\mu}, \tau) \sim \frac{c_{N-1}}{c_N} \frac{|\tau|^\nu}{w_\nu(\hat{\mu}, -\tau^2)} R_{N_f=0,\nu}^{\mathbb{R}}(\hat{\mu}; -\tau^2)\,. \tag{A.45}$$

In the weak limit, it follows from (3.138) that

$$\lim_{N\to\infty,\,\mathrm{weak}} R_{N_f=0,\nu}^{\mathbb{R}}(\hat{\mu}; -\tau^2) \sim \rho_w^{(N_f=0,\mathbb{R})}(\gamma; -\kappa^2) = -G_w(-\kappa^2, -\kappa^2)\,, \tag{A.46}$$

$$\lim_{N\to\infty,\,\mathrm{weak}} \frac{|\tau|^\nu}{w_\nu(\hat{\mu}, -\tau^2)} = \frac{1}{\hat{h}_w(-\kappa^2)}\,, \tag{A.47}$$

where $\hat{h}_w$ is defined in (3.140). From (3.98), we have

$$\lim_{N\to\infty,\,\mathrm{weak}} \frac{c_{N-1}}{c_N} \sim \frac{(2\hat{\mu})^{-(N-1)(N-1+\nu)}\eta_+^{-(N-1)(N+\nu-2)/2}}{(2\hat{\mu})^{-N(N+\nu)}\eta_+^{-N(N+\nu-1)/2}} = (2\hat{\mu})^{2N+\nu-1}\eta_+^{N-1+\nu/2} \tag{A.48}$$

$$\sim \sqrt{2N}\hat{\mu}\,\mathrm{e}^{N\hat{\mu}^2} = \gamma \mathrm{e}^{\gamma^2/2}\,. \tag{A.49}$$

Collecting all results, we finally obtain (3.158),

$$\mathcal{Z}_w^{(N_f=\|1\|,\,\nu)}(\gamma, \kappa) \sim -\gamma \mathrm{e}^{\gamma^2/2} \frac{G_w(-\kappa^2, -\kappa^2)}{\hat{h}_w(-\kappa^2)}\,.$$

### A.4.4 The Large-N Limit of Maximal Non-Hermiticity

Below we derive an analytical expression for

$$\mathcal{Z}_s^{(N_f=\|2\|,\,\nu)}(\tau_1, \tau_2) \equiv \lim_{N\to\infty,\,\mathrm{maximal}} \left\langle |\det[\mathcal{D}(\hat{\mu}) + \tau_1]\det[\mathcal{D}(\hat{\mu}) + \tau_2]| \right\rangle_{N_f=0}\,, \tag{A.50}$$

where $N$ goes to infinity with $\hat{\mu} = 1$ and $\tau_f$ fixed.

Putting $\hat{\mu} = 1, k = 2, N_f = 0, x_1 = -\tau_1^2$, and $x_2 = -\tau_2^2$ in (A.40) we find

$$Z_{N_f=\|2\|,\nu}^{\mathrm{RMT}(N-2)}(1, \{\tau_1, \tau_2\}) \sim \frac{c_{N-2}}{c_N} \frac{1}{|\tau_1^2 - \tau_2^2|} \frac{|\tau_1\tau_2|^\nu}{w_\nu(1, -\tau_1^2) w_\nu(1, -\tau_2^2)} R_{N_f=0,\nu}^{\mathbb{R}}(1; -\tau_1^2, -\tau_2^2). \tag{A.51}$$

The factor $c_{N-2}/c_N$ is independent of masses and may be dropped. According to [4, Eq. (2.18)], the quenched two-point function for real squared Dirac eigenvalues reads

$$R^{\mathbb{R}}_{N_f=0,\,\nu}(1;x_1,x_2) = w_\nu(1,x_1)w_\nu(1,x_2)f_N(x_1,x_2), \tag{A.52}$$

with

$$\begin{aligned} f_N(x_1,x_2) \equiv {} & \mathcal{K}_{N,\nu}(1;x_1,x_2)\,\mathrm{sgn}(x_1-x_2) \\ & + \int_{\mathbb{R}} dx \int_{\mathbb{R}} dx' w_\nu(1,x)w_\nu(1,x')\Big[\mathrm{sgn}(x_1-x)\,\mathrm{sgn}(x_2-x') - \mathrm{sgn}(x_2-x)\,\mathrm{sgn}(x_1-x')\Big] \\ & \times \Big[\mathcal{K}_{N,\nu}(1;x_1,x)\mathcal{K}_{N,\nu}(1;x_2,x') - \frac{1}{2}\mathcal{K}_{N,\nu}(1;x_1,x_2)\mathcal{K}_{N,\nu}(1;x,x')\Big], \end{aligned} \tag{A.53}$$

with $\mathcal{K}_{N,\nu}$ given in [4, Eq. (2.20)] explicitly. We plug (A.52) into (A.51) to get

$$\mathcal{Z}_s^{(N_f=\|2\|,\,\nu)}(\tau_1,\tau_2) \sim \frac{|\tau_1\tau_2|^\nu}{|\tau_1^2-\tau_2^2|} f_\infty(-\tau_1^2,-\tau_2^2),$$

where $f_\infty(x_1,x_2)$, defined as the large-$N$ limit of $f_N(x_1,x_2)$, can be computed with the aid of the strong limit of $\mathcal{K}_{N,\nu}$ given in [4, Eq. (2.30)]:

$$\lim_{N\to\infty} \mathcal{K}_{N,\nu}(1;u,v) = \frac{1}{64\pi}(uv)^{-\nu/2}(u-v)I_\nu(\sqrt{uv})\,. \tag{A.54}$$

This completes the derivation of $\mathcal{Z}_s^{(N_f=\|2\|,\,\nu)}(\tau_1,\tau_2)$ .

## References

1. T. Schäfer, Mass terms in effective theories of high density quark matter. Phys. Rev. **D65**, 074006 (2002)
2. T. Schäfer, QCD and the eta' mass: instantons or confinement? Phys. Rev. **D67**, 074502 (2003)
3. N. Yamamoto, T. Kanazawa, Dense QCD in a finite volume. Phys. Rev. Lett. **103**, 032001 (2009)
4. G. Akemann, M.J. Phillips, H.J. Sommers, The chiral Gaussian two-matrix ensemble of real asymmetric matrices. J. Phys. **A43**, 085211 (2010)

MIX
Papier aus verantwortungsvollen Quellen
Paper from responsible sources
FSC® C105338

If you have any concerns about our products,
you can contact us on
**ProductSafety@springernature.com**

In case Publisher is established outside the EU,
the EU authorized representative is:
**Springer Nature Customer Service Center GmbH**
**Europaplatz 3, 69115 Heidelberg, Germany**

Printed by Libri Plureos GmbH
in Hamburg, Germany